T0210888

Lecture Notes in Computer Science 9990

Commenced Publication in 1973
Founding and Former Series Editors:
Gerhard Goos, Juris Hartmanis, and Jan van Leeuwen

More information about this series at http://www.springer.com/series/8851

Ngoc Thanh Nguyen · Ryszard Kowalczyk
Cezary Orłowski · Artur Ziółkowski (Eds.)

Transactions on Computational Collective Intelligence XXV

 Springer

Editors-in-Chief

Ngoc Thanh Nguyen
Department of Information Systems
Wrocław University of Technology
Wroclaw
Poland

Ryszard Kowalczyk
Swinburne University of Technology
Hawthorn, VIC
Australia

Guest Editors

Cezary Orłowski
Gdansk School of Banking (WSB Gdańsk)
Gdańsk
Poland

Artur Ziółkowski
Gdansk School of Banking (WSB Gdańsk)
Gdańsk
Poland

ISSN 0302-9743   ISSN 1611-3349   (electronic)
Lecture Notes in Computer Science
ISBN 978-3-662-53579-0   ISBN 978-3-662-53580-6   (eBook)
DOI 10.1007/978-3-662-53580-6

Library of Congress Control Number: 2016953315

This Springer imprint is published by Springer Nature
The registered company is Springer-Verlag GmbH Germany
The registered company address is: Heidelberger Platz 3, 14197 Berlin, Germany

Transactions on Computational Collective Intelligence XXV

Preface

Modern agglomerations face the challenge of changes arising from the needs and requirements of their residents, and from either acceptance or rejection of the "smart cities" vision. The consideration of these requirements and the acceptance of the vision are a long-term process in which municipal decision-makers, city residents, and civic organizations work out a compromise, which is often the result of merit-based decisions by the authorities but can also result from political decisions on which the residents only have an indirect influence. Such a complex city system – seen from the perspective of the authorities, city residents, and organizations, and taking into account many decision-making processes that are hard to control and analyze – represents a complex environment for the implementation of information technology supporting city management processes.

Owing to the aforementioned considerations, the process of IT implementation represents a system of complex technology- and management-related mechanisms (more focused on management-related ones), whose pre-implementation analysis becomes crucial for building a successful strategy for the completion of such projects. Therefore, a relatively large amount of information is published on the functioning of cities in the context of their transformation to smart cities and on the technologies applied both in system design and implementation; experiences are also presented of cities that were, have been, or will be in some stage of such a transformation.

The set of papers presented here (prepared by the team) is part of the presentation of descriptions of such transformation processes. It is based on the experiences of the CAS design team (IBM Centre for Advanced Studies on Campus), making use of IBM IOC (Intelligent Operating Centre) consisting of six members (Cezary Orłowski, Tomasz Sitek, Artur Ziółkowski, Paweł Kapłański, Aleksander Orłowski, Witold Pokrzywnicki). The technology framework of smart cities systems (where IOC may be given as an example) shows the opportunities and constraints for the implementation of city processes. It also enables a broader model-based analytical view of city processes and the specific information technologies applied in order to model and implement these processes. Taking into account this form of presentation, the papers consider three perspectives of the design and implementation of smart cities systems.

The first perspective is the client perspective, i.e., of the city and its organizational processes and the possibilities of applying measurements to these processes. In the first paper, "High-Level Model for the Design of KPIs for Smart Cities Systems," two points of view are considered: a high-level view within which the city processes are discussed and confronted with measurements in the form of key performance indicators (KPIs) and a low-level one showing to what degree the available indicators may be applied to measure the city's processes. Within this perspective in the second paper, "Implementation of Business Processes in Smart Cities Technology," the model of the city processes is presented and the authors' own measurements for assessing the maturity of these processes are suggested. Moreover, opportunities for enhancing the KPIs through creating integrated or dynamic KPIs are indicated. These two papers aim (a) at showing to what extent the present approaches based on KPIs may be applied in the design framework delivered by software developers and (b) at suggesting measurements for assessing the maturity of these processes.

The second perspective is the project perspective, on which two papers are presented. In the paper "Designing Aggregate KPIs as a Method of Implementing Decision-Making Processes in the Management of Smart Cities," a low-level view of the project in the context of management processes is described. The fourth paper, "Smart Cities System Design Method Based on Case Based Reasoning," illustrates an approach resulting from the need to treat both the development process management method and system implementation as components that may be used by any city. Both of these papers provide methodology-based support for the management and implementation processes of smart cities systems.

The third perspective is the provider's perspective. Here, two papers are presented that describe low-level and high-level approaches. In the fifth paper of the volume, "Model of an Integration Bus of Data and Ontologies of Smart Cities Processes," the high-level approach to using an ontology for supporting the construction of a high-level architecture is presented. The construction of such an architecture becomes necessary in the case of an agile approach to project management. The authors' experiences connected with use of agile methods show that the availability of an ontology of concepts (objects and processes, both development-related and management-related ones) significantly simplifies the design of sprints and the prioritizing of backlog tasks. In the sixth paper, "Ontology of the Design Pattern Language for Smart Cities Systems," the second low-level perspective, the significance of building an integration bus for a joint view of development processes, technology, and artifacts, as well as the products of the design and implementation of smart cities are described.

Additionally, we include two papers concerning the dynamic and semantic assessment of systems. In their contribution, Vo Thanh Vinh and Duong Tuan Anh propose two novel improvements for minimum description length-based semisupervised classification of time series: an improvement technique for the minimum description length-based stopping criterion and a refinement step to make the classifier more accurate. In the eighth paper by B. John Oommen, Richard Khour, and Aron Schmidt, the problem of text classification is explained using "anti"-Bayesian quantile statistics-based classifiers.

The papers presented are the result of shared projects on organizational solutions, carried out together with IBM, such as the 10-year period of collaboration within the Academic Initiative, Competence Centre and Centre for Advances Studies on Campus, and also research projects carried out at the Gdańsk University of Technology and CAS. During 2011–2015, the international research project Eureka E! 3266 (EURO-ENVIRON WEBAIR) "Managing Air Quality in Agglomerations with the Use of a www Server" was carried out. The Armaag Foundation, IBM, DGT, Gdańsk City Council, and the Marshall's Office in Gdańsk all took part in the project. The project objective was to create an IT system supporting decisions with regard to dust pollution and noise in Gdańsk. Hence the project was addressed to City Council analytical units, which deal with the conditions of such decisions.

The second project was the PEOPLE MARIE CURIE ACTIONS project carried out within the International Research Staff Exchange Scheme called: FP7-PEOPLE-2009-IRSES "Smart Multipurpose Knowledge Administration Environment for Intelligent Decision Support Systems Development," and continued until the end of March 2015. The goal of the project was the development by the Australian partner (University of Newcastle) of an environment for the building of intelligent decision support systems based on SOEKS (Set of Experiences). The data/cases for the verification of the environment were provided by the partners, namely, the Gdańsk University of Technology and Vicomtech from Spain. In the schedule of the project, three verification cases had been envisaged, and one of them was the data concerning the design of a smart cities system for Gdańsk within the Eureka project.

The synergy of these two projects and the experience of many business partners collaborating in both projects, as well as the close cooperation between CAS and IBM Polska, created the conditions for such a comprehensive assessment of smart cities systems. The three perspectives presented in the work – i.e., that of the client of the city, the smart cities for the Gdańsk project, and the provider, CAS Gdańsk – close the first stage of experiences covering system design and implementation. The papers on this work (covering the three perspectives) were prepared so as to have a generic and component-specific dimension and may serve as guidelines in both the design and implementation of smart cities systems for a number of cities.

September 2016   Cezary Orłowski
                                                                 Artur Ziółkowski

Transactions on Computational Collective Intelligence

This Springer journal focuses on research in applications of the computer-based methods of computational collective intelligence (CCI) and their applications in a wide range of fields such as the Semantic Web, social networks, and multi-agent systems. It aims to provide a forum for the presentation of scientific research and technological achievements accomplished by the international community.

The topics addressed by this journal include all solutions to real-life problems for which it is necessary to use computational collective intelligence technologies to achieve effective results. The emphasis of the papers published is on novel and original research and technological advancements. Special features on specific topics are welcome.

Editor-in-Chief

Ngoc Thanh Nguyen   Wroclaw University of Science and Technology,
                           Poland

Co-Editor-in-Chief

Ryszard Kowalczyk   Swinburne University of Technology, Australia

Guest Editors

Cezary Orłowski   Gdansk School of Banking (WSB Gdańsk), Poland
Artur Ziółkowski   Gdansk School of Banking (WSB Gdańsk), Poland

Editorial Board

John Breslin   National University of Ireland, Galway, Ireland
Longbing Cao   University of Technology Sydney, Australia
Shi-Kuo Chang   University of Pittsburgh, USA
Oscar Cordon   European Centre for Soft Computing, Spain
Tzung-Pei Hong   National University of Kaohsiung, Taiwan
Gordan Jezic   University of Zagreb, Croatia
Piotr Jędrzejowicz   Gdynia Maritime University, Poland
Kang-Huyn Jo   University of Ulsan, Korea
Yiannis Kompatsiaris   Centre for Research and Technology Hellas, Greece
Jozef Korbicz   University of Zielona Gora, Poland
Hoai An Le Thi   Lorraine University, France
Pierre Lévy   University of Ottawa, Canada
Tokuro Matsuo   Yamagata University, Japan
Kazumi Nakamatsu   University of Hyogo, Japan
Toyoaki Nishida   Kyoto University, Japan

Contents

High-Level Model for the Design of KPIs
for Smart Cities Systems

Cezary Orłowski[1(✉)], Artur Ziółkowski[1], Aleksander Orłowski[2],
Paweł Kapłański[2], Tomasz Sitek[3], and Witold Pokrzywnicki[3]

[1] WSB University in Gdańsk, Gdańsk, Poland
{corlowski,aziolkowski}@wsb.gda.pl
[2] Gdansk University of Technology, Gdańsk, Poland
{aleksander.orlowski,pawel.kaplanski}@zie.pg.gda.pl
[3] Staples Advantage Poland SP. Z O.O., Gdańsk, Poland
tomasz.sitek@staples.com,
witold@urtico.com

Abstract. The main goal of the paper is to build a high-level model for the design of KPIs. Currently, the development and processes of cities have been checked by KPI indicators. The authors realized that there is a limited usability of KPIs for both the users and IT specialists who are preparing them. Another observation was that the process of the implementation of Smart Cities systems is very complicated. Due to this the concept of a trigger for organizational-technological changes in the design and implementation of Smart Cities was proposed. A dedicated Model for City Development (MCD) was presented. The paper consists of four main parts. First the structures of both city and business organizations were presented. Based on that, in the second part, the processes existing in cities and business organizations were presented to show how different they are. The third part presents the role of KPIs and their limitations with the example of the IOC. The last part consists of the presentation of the model and its verification based on two city decision-making examples. The proposed design model presented herein takes into account both the city indicators and their aggregate versions for the needs of city models.

Keywords: Smart cites · Knowledge base · Knowledge management · Fuzzy logic · Process modeling · Decision support

1 Introduction

Currently, 54 % of the people who live in the world live in city areas. According to the United Nations this is 3,5 billion people and is supposed to grow to 7 billion in 2045 [1]. The process might be most visible especially in North America (84 % of the population living in urban areas) and Europe (73 %).

The data shows that managing city areas is, and is going to be, more important with the growing number of inhabitants and limited area in which the cities might and should (e.g. because of economic reasons) grow. It should be noted that it is not only a process of fast-growing cities, there are many examples (with the best-known: Detroit), where the number of inhabitants is rapidly dropping. In both cases pure managerial

© Springer-Verlag GmbH Germany 2016
N.T. Nguyen et al. (Eds.): TCCI XXV, LNCS 9990, pp. 1–14, 2016.
DOI: 10.1007/978-3-662-53580-6_1

decisions have to be taken. There is a need to make managerial decisions but it is not obvious what kind of decisions are going to be taken because a city is a type of organization in which different attitudes to the same problem are seen. This might be illustrated by one of the typical problems: should a city build more highways in the city centers, which is very expensive for the city and devastates the surrounding areas plus increases congestion but is expected by local inhabitants and therefore has a strong political influence?

If there is a need to manage cities more effectively because of the growing number of inhabitants and there are no clear rules/values according to which decisions are taken it seems to be necessary to help cities in the process of effective management. How this can be done and what kind of limitations are observed will be presented in this paper.

2   A City and a Business Organization

A city and a business organization are two types of organization. An organization is a "formalized intentional structure of roles and positions [2]." Although they are types of organization there are several differences between them and in managing them because "Management applies to any kind of organization [2]." These aspects will be presented below.

A business organization (also known as an enterprise) is an entity formed for the purpose of carrying on commercial enterprise. Such an organization is based on systems of law governing contract and exchange, property rights, and incorporation [3]. Management was originally dedicated to business organizations starting from Henry Fayol and Fredric Winslow Taylor at the end of XIX centry. The main goal of a business organization is to generate surplus. While managing the enterprise the interest of owners/shareholders, employees and business partners should be taken into consideration.

The enterprise might offer products or services as the main result and both might be offered either to individuals (B2C) or business customers (B2B). Most current knowledge in management is concentrated on managing business organizations with dedicated models supporting that process, special indicators used for this and many scientific methodologies. Because this is so obvious it will not be presented in this paper.

A city is "an inhabited place of greater size, population, or importance than a town or village [4]." "Cities should be seen in terms of networks stretching in time and space [5]." There are many different attempts at the definition of a city. Here, one might present the idea of "The Ideal-Type City" by Max Weber or the IRN (Inter-Representation Network) Cities.

There are several other descriptions of a city, "The city is a network of networks, embedded in broader networks, and within it are the values flows between network participants [6]." A city can be presented as a social area but also a physically existing area, and it is often also described as a cultural area with socio-economic processes.

The process of analyzing a city from the managerial point of view has a long history. In 1970 the NYC-RAND Institute used urban statistics, modeling and computation developed for wartime and typical corporate management to determine resource allocation, especially for New York City's Fire Department [7]. After more than 40 years the question should be asked, will new technologies help manage cities in a better way? To answer that it seems to be necessary to present the procedures that should be supported by technologies in the current cities.

City management is normally seen from two perspectives: managing the city hall and managing the whole city. Managing the whole city consists of aspects such as city strategy creation, and the control, coordination and assessment of departments which are implementing the city's strategy and policy.

The result of city management is a city product which is different from a typical commercial product offered by enterprises. The main differences between these two types of products consist of:

- High complexity of the city product
- Limited market influence on the product
- Consumption of the product in one, defined, location
- Very complicated process of pricing parts of the product (social climate, image) which strongly influences how attractive the product is [8].

It is important to mention that today's developed cities 'are social and technical complex systems characterized by historically unprecedented levels of diversity and temporal and functional integration [7]." There is a growing individual specialization and interdependence which makes large citie 'extremely diverse and crucially relies on fine temporal and spatial integration and on faster and more reliable information flows.' Because of that cities are the economic and cultural engines of all human societies."

3 Types of Processes in a City vs Processes in a Company

As it was presented in previous paragraphs a city and a company while both being types of organization are defined differently. It also means that the processes which are used in both city management and managing a company are different. In this section the differences will be presented.

As a result of every process it is believed that the better it fulfills the process requirements the better the whole organization exists. However, when cities are discussed, an important statement should be referred to: "The world's most vibrant and attractive cities are not usually the same places where buses run impeccably on time. While improvements in infrastructure and urban services are absolutely necessary for cities to function better, they are not the fundamental sources of social development or economic growth [7]."

Current cities, as described above, are unique examples of organizations. Besides the description, it seems to be important to present how cities work, how they are organized and what type of processes might be observed.

Typical cities in Poland are divided into departments. A department is "a distinct area, division, or branch of an organization over which a manager has authority for the performance of specified activities." While discussing business organizations there might be several types of departments presented (e.g. sales department, production department). Companies try to adapt departmentalization into their main type of business: it might be departmentalization by time (when shifts are in use), departmentmentalization by geography (when a company tries to adapt to several markets in dfferent geographic locations), customer departmentalization (when different types of customers seem important to be represented in organization structure) or several others. When taking cities into consideration they are typically divided by enterprise functions so there are departments representing functions of the city such as financial department, architecture and urbanistic department, social department, etc. Such departmentalization has several advantages presented in the theory of organization from which it is worth mentioning that it follows the principle of job specialization, simplifies training and seems to be logical and, because of this, easy to create that type of structure.

This type of departmentalization has also disadvantages, from which the most important is that people working in one department have problems with seeing the organization as a whole. It creates 'walls' between departments; employees mostly do not know what is done in other departments.

It is also important to mention wicked problems which are found in city management. The essential character of wicked problems is that they cannot be solved in practice by a central planner. This is based on two types of problems (a) the knowledge problem (b) the calculation problem [7]." Calculation can be easily done by today's computers but 'the *knowledge* problem refers to the information that a planner would need to map and understand the current state of the system; the city, in our case. While still implausible, it is not impossible to conceive information and communication technologies that would give a planner, sitting in a 'situation room', access to detailed information about every aspect of the infrastructure, services and social lives in a city. Privacy concerns aside, it is conceivable that the lives and physical infrastructure of a large city could be adequately sensed in several million places at fine temporal rates, producing large but potentially manageable rates of information flow by current technology standards' which will mean it is not a problem in city management.

It is also necessary to define when it is possible to conclude that a city is properly managed. In a business organization the results are normally presented in crisp values (such as a financial result at the end of the fiscal year, the growth/drop in the number of sold products, the growth of stock value). When the same question is posed for cities the answer is not clear. From the city mayor's perspective the main indicator of city management is the result of the election. But it cannot be concluded that the results are based only on crisp results of the city (such as city debt, level of infrastructure development, etc.). It is often based only on feelings or personal opinions which might be (and many times are) very different from real results.

Even if the assumption can be made that city managers will not follow political needs (to win the election) but will concentrate on the needs of the city, a similar question might be posed – what are the needs of the city? The needs are defined by different actors (inhabitants, investors, politicians, public organizations). One of the suggestions of how to answer that question is the idea of Smart Cities.

The Smart City notion is a concept that started to emerge approximately two decades ago and was originally used to describe a city that applied technological solutions to the everyday problems of the city and its inhabitants, through the intensive use of information and technologies [9]. This can be prested as a definition of smart cities but the question arises as to when the current existing city might be called a smart city. "...when investments in human and social capital, transport and ICT fuel sustainable economic growth and a high quality of life, with a wise management of natural resources, through participative governance [10]". There are also more general definitions as the one from the European Smart Cities Model that states that a Smart City is a city that performs well in six areas: Economy, Mobility, Environment, People, Quality of Living and Governance [11]. So a Smart City is more than an intelligent city because it creates and uses feedback.

To answer the question of 'how to manage effectively' it is necessary to present the type of methodologies that might support decision making. It seems to be necessary to refer here to the concept of Smart Cities (presented above). These days the Smart Cities 2.0 concept is becoming popular as an idea in which departments are connected through digital strategies which helps to integrate and build bridges between the current 'silos' (represented as different departments) [12].

Because a city cannot be seen as the same type of organization as a company (business organization) it means that it should be managed also in a different way. As presented in previous paragraphs, it has different goals and because of this another logic of existence and the way is it managed. It also means that different tools for supporting management processes should be used. Even if the logic and goals are different and the whole process of managing the city is more complicated than in a typical business organization, it still has to be supported.

3.1 Examples of City Management Processes

As it was presented in the previous section, cities are divided into partly independent departments. This presents just the organizational structure which by itself should not impact directly on city management. It is important to present how different processes organized by each department influence the process of city management. Next, the examples ill be presented.

The process of investing money in the transport infrastructure influences several different areas such as the location of schools, land value, pollution and economic development. One of the major decision-making problems of this kind is the project of Podwale Przedmiejskie Street in Gdansk. Today, Podwale Przedmiejskie Street is seen as a major spatial barrier and a burden to its neighbours. It was built 40 years ago as a transit road through the historical city centre and consequently divided the city. As an effect, one side is still perceived as a high quality district, a popular tourist destination with all the famous landmarks, and functions as a city centre. The other side, however, is considered a dangerous, impoverished district even though it has a lot of valuable, historical urban tissue which - unlike most of the rest of the city - survived the 2nd World War. Because of this there is an idea to rebuild the street, narrow the whole street and build a pedestrian crossing. There are several sides interested in this topic: car owners are against it, city and non-profit organizations hope that it will 'bring back to life' a huge part of the city, and several other factors from different areas have to be taken into consideration (like air pollution, noise but also access to shops and services located in this area and the total cost of constructions). These areas cover the interest of several city departments and in the end the city, as a whole, has to make the decision.

The key question is who is going to assess the influence on the different areas and how to calculate the final results from several areas.

Another case is the changes in the network and location of schools in the city. The changes are normally organised by the department of education and are mostly based on data such as demography in the neighbourhood, the market needs, the school facilities (gym, swimming pool). Based on this kind of data, decisions about closing, opening new or relocating schools are made. Several examples might be presented here but one of the most noticeable was a few years ago in one of the cities in Poland. There was a relocation of schools and slight changes in the school timetable made by the department of education. At the same time, the department of transportation noticed high changes in the transportation system of the city (congestion on some bus lines whereas others rapidly became empty) and traffic jams in new places and times. Special research was done and based on the results the department realised that there were changes in the school network made by other departments of the same city.

The examples described above present the main problem existing in the current process of city management – how to support decision-making processes in the whole city (where normally decisions are made at the level of departments).

To answer that question it is necessary to define the proper way of city development. It was defined in the first section of this paper. When it is known in what direction the city should go it is possible to think of how it might be done.

4   Key Performance Indicators and Their Significance for Estimating Cit Processes in IOC

In the previous sections the problems of managing cities were presented. Based on the examples, current trends and main problems, it seems to be important from one side, but also complicated, to manage cities efficiently. Besides managing cities, it is important to find a technology that might support that process. The last great technological advancement that reshaped cities was the automobile (and the second in importance was the elevator). In both cases, these technologies reshaped the physical aspects of living in cities – how far a person could travel or how high a building could be. But it did not change the fundaments of the city because it was connected only with technology. Currently, when personal computers, mobile phones and the Internet are in use, there is the ability to influence also the social organization of cities and empower everyday citizens with the knowledge and tools to actively participate in the policy, planning and management of cities [13]. This is what the Smart Cities concept tries to use. Besides having just a concept there is a need to have tools that might be practically used by cities. An example of these tools, which is going to be presented next, is IBM's Intelligent Operations Center for Smarter Cities (IOC).

The IOC is able to receive, transform and use the data gathered from many different sources to support city management processes. It is a big system (big data) that consists of a lot of features and extensions.

The main element of the IOC are Key Performance Indicators (KPIs). A KPI "is a measurable value that demonstrates how effectively a company is achieving key business objectives. Organizations use KPIto evaluate their success at reaching targets [14]."

KPIs include Data Source, Model and KPI itself. They help an organization define and measure progress towards organizational goals. Here the key question should appear: what is the main goal of the city?

Several answers might be given here:

- Build 2 new roads
- Reduce the unemployment rate in the city by x percent
- Build 3 new schools

This reflects the discussion presented in the first part of this paper in trying to answer the question of what a successful city looks like. Based on the current knowledge, it is a mix of socio-economic aspects. Here the second feature of KPIs should be presented: every KPI must be measurable. Several examples of KPIs used in business might be presented:

- A business may have as one of its Key Performance Indicators the percentage of its income that comes from return customers.
- A Customer Service Department may have it as a percentage of customer calls answered in the first minute [15].

KPIs are mostly used in managing organizations but it seems to be important to check if they can be used for city management. In systems like the IOC hundreds of KPIs should be taken into consideration. In current cities the amount of collected data is significant. Based on this the KPIs are built and might be presented to the system operators. But still this concentrates only on measurable crisp values, which skips many socio-economic aspects very important for the city.

The authors prepared two types of models: a model which will help in organizing the KPIs in the IOC (by dividing the KPIs into categories) and a model of city management processes (MCMP) which will present another view on the problem analysis in the IOC.

5　High-Level Model for the Design of KPIs for Smart Cities

The starting point for the construction of a high-level model for the design of indicators (WMPW) was to assess the management processes of cities and organizations described in the previous section. It was assumed that this differentiation in the processes of a city and an organization shows a limited application and design of KPIs with a bottom-up approach. Also the integrated KPI models that are presented in a different submitted paper indicate that they can be designed in conditions in which while managing a team of designers one is aware of a high-level use of indicators. While the paper entitled "Designing aggregate KPIs as a method of implementing decision-making processes in the management of Smart Cities" discusses the design of aggregate indicators with a view to the aggregation of these indicators, this work discuss the need for the design of indicators with the model of city processes at its basis. The research presented in this paper also shows that the indicators can/should be designed from the top according to a top-down approach unlike that presented in the previously quoted paper. These two different approaches can be used depending on the maturity of the project team and the representatives of cities.

In a situation in which this maturity is high, the indicator design process using a bottom-up method appears to be more efficient. However, when taking a top-down approach, the maturity of the team may be low, provided, however, that both teams are familiar with detailed models of the city while building the metaphor of the system.

Therefore, this paper proposes an extended approach to building KPIs based on models depicting how the city functions. It was assumed that the adoption of such a model for implementation forms the basis for the design of indicators as well as for their integration. It was also assumed that the adoption of such a model also acts as a metaphor of the (constantly developing) city processes, readable for users of city systems and also for system designers. Hence, the development of a high-levl model for the design of indicators (WMPK) may provide a kind of trigger for changes in the organization of cities and in the method of evaluating processes and their importance for decision making.

The starting point for the design of indicators was the analysis of models of city processes. Indeed it was assumed that the scope and number of these models will indicate to what extent the approach to these models is important or integrated (attempt at their aggregation) or to treat these models as independent entities and use them as design patterns on the basis of defined KPIs. The top-dow approach was deliberately used to indicate the importance of knowing the vision of the operation processes of the city before the defining of indicators for this vision. This approach is commonly used in the design processes of corporate architectures and can constitute a methodological component used in the design of KPIs.

The analysis of city processes indicates that the number of models of the operation processes of cities is limited. City processes and the need for their use and credibility in the design of Smart Cities systems indicates that the suitability of these models for KPI design processes must be evaluated in order to then generalize this process to assume an approach under which initially (along with the city) a model of the operation processes of the city is adopted and then KPIs are designed bearing in mind the possibility of their aggregation for the needs of this model.

Figure 1 presents WMPW where there are three visible layers (city models, aggregate indicators and KPIs, which are the basis for measuring those processes of the city that are important from the point of view of city models). The feedback vector visible in this figure and the controller on the right-hand side indicates the direction and area of the design processes. This city model represents specific kind of data necessary for the design of indicators. Based on the city models, processes are selected and measurements are assigned to these processes. Then the aggregation of indicators takes place based on the processes isolated within the models and they are assigned to the corresponding measurements.

Because of the situation presented above there are currently thousands of KPIs in the IOC system. Some of these are pure business KPIs (which cannot be used in the city management process), the others might be used, all are put together and the new KPIs are added in the same way. The authors propose to add extra levels to the extent in which all data representation in the IOC is based on KPIs. It is suggested to first to add aggregate KPIs which will accumulate KPIs in order to meet important issues.

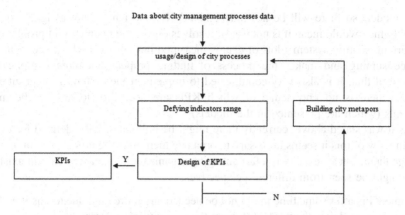

Fig. 1. High-level KPI design model (WMPW) for the needs of smart cities

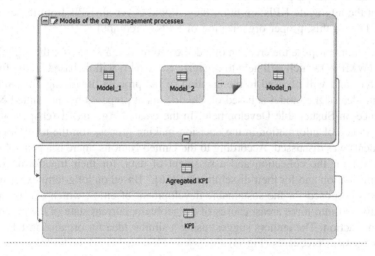

Fig. 2. Three layer architecture of the KPI design (additional layer of the city project management model)

This idea is presented in detail in the next paper and therefore will not be presented here. The authors suggest adding one more level: models of city management processes (Fig. 2), which will be presented further.

As presented in Fig. 2 the top part is the model of city management processes. The model represents the main idea to be implemented in the city (for example a model for sustainable development or a model for effective transportation). So first the model is described. Next to that model aggregate KPIs are assigned which will represent the main areas of interest of the model (such as all important aspects of transportation in the transportation model). The aggregate KPIs consist of many individual KPIs necessary to describe the model. Every KPI that will be created will be assigned to one or many models so there will be no 'independent' KPIs which are not assigned to any model (which would mean it is not used). It solves one of the main current problems in the current existing systems: lots of data is measured but not used which is expensive, time-consuming and makes the process of finding proper data more complicated. Because of that, it is also very complicated to implement the system in other cities – because there is no knowledge of which KPIs are necessary to answer the main problems (which are presented in the models).

As it was stated above, currently there might be billions of KPIs defined for every city. In view of this it seems necessary to organize them to make it possible not only to manage them. Because a city is a fast changing organization, managing current existing KPIs might be seen from different perspectives:

- Some KPIs after some time might not be needed any more (and due to this it will be necessary to delete them which lowers the cost and gives order.
- When a new decision-making process is going to be made, very often it is possible to base it on current existing KPIs. As there are very many of them in the city, it is required to make it possible to find them.
- When the aggregate KPIs are built, they should cover all KPIs from the necessary area. Due to this, proper organization of KPIs is required.

The authors propose the creation of a dedicated model, a model for the organization of KPIs (WMPW) which will help to organize KPIs. This will be based on the function of eery KPI and will be used in the creation of the process of solving city problems. The main idea of the model is based on the document prepared by the United Nations Conference on Sustainable Development. In the created Agenda 21 (chapter 40) [16] the importance of information in the decision-making process (on the level of country government) was discussed. According to the United Nations 'there is a general lack of capacity (..) for the collection and assessment of data, for their transformation into useful information and for their dissemination [16]." Based on long-term experiments a model was proposed for the sustainable development of cities, which consists of 130 factors divided into three areas: causes of the problem, current state of the process and proposed reaction. The authors suggest using a similar idea for organizing KPIs in the city management processes (Fig. 3).

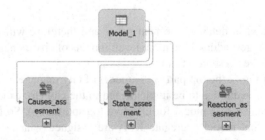

Fig. 3. Model of the city management process

The main idea of the model for the organization of KPIs is to solve the problems presented above. In this respect the model is divided into three parts: causes, state and reaction.

The proposed construction will not only help organizing the KPIs (their place in the system) but mostly should help in the main process for which KPIs are used: building the procedure in the decision-making process. As presented in the first section of the paper, there are different needs placed by different groups in cities. The first level of the model (Causes) will help in defining the potential needs/problems proposed by different groups in the city. When the needs are seen (and measured, for example, how important the need is) the KPIs from Causes will lead us to the State part in which the KPIs presenting different areas of the city are presented. Here all the everyday measures are presented (e.g. traffic, pollution, budget). For the declared needs one (or a group) of KPIs will reflect the current state of the city in the area in which the problem might occur. This will lead the user to the third part called Reaction. It will consist of the KPIs that will measure the potential reaction of the city to the presented problem, taking into consideration the current state.

It is suggested that:

- 'Causes' consists of all KPIs which are defined as people's needs; pressure on several processes that should happen according to people's beliefs. These KPIs will measure those needs.
- 'State' consists of KPIs presenting the current values of processes (all measured in the city).
- Reaction should consist of the reaction to negative trends that might appear, also for that, KPIs are necessary.

The proposed construction of the model supports the model of city management processes presented in the first part of this section. The presented organization of KPIs supports the process of building aggregate KPIs because the potential user can easily find the KPIs needed for the aggregation. It also shows that KPIs do not have to be described (and assigned) to the categories based only on the area in which they exist (such as management, pollution, transportation).

6 Verification of City Models

In the previous section the idea of the model was presented. This section will present the verification of the presented model.

The usage of the model will be verified on the example of a model built for sustainable development for Warsaw. Therefore, it is the model for sustainable development for Warsaw that consists of four aggregate KPIs (urban/environment, economic, social, management/political). Each of the aggregate KPIs consists of a certain number of KPIs. For better understanding there was a sub-category added and called Area (each aggregate KPI consists of a certain number of KPIs which are grouped into Areas in view of their main goal).

Table 1 present the sustainability of the city there must be in total 297 KPIs taken into consideration. These represent several different areas. Keeping in mind that this is

Table 1. Aggregate KPIs, areas & number of KPIs

Aggregate KPI	Area	Number of KPIs
Urban/environment	Water management Waste water management Rubbish management Green areas management Land management	74
Economic	Number of companies/unemployment Structure of companies Availability of services/media agriculture	42
Social	Demography Labor market Housing Culture and tourism Education and science Environmental protection	73
Management/political	Management Budget (income) Budget (expenses)	108
Total		297

just one model (out of many others in the city) it shows the scale of usage of KPIs in the city. Because every KPI is assigned to a model in which it is used, it is possible to verify and maintain the usage of every KPI (which also means the need to monitor the factors used in each KPI). It helps to avoid the situation in which there are KPIs which are not used at all (not used in any model). Even more important seems to be the fact that because of the structure it is possible to easier implement the IOC in another city – copying the model means that there is a list of necessary KPIs to be used to make it possible to receive the necessary model.

7 Conslusions

This paper presents a high-level model of the design of indicators for smart cities. The starting point for their design was the negative experience of the authors in the design of indicators for the evaluation of individual processes of city management. It was proposed to use, in the design of indicators, city operating models for which areas for the aggregation of indicators are determined in order to, on this basis, design individual indicators for city processes.

The paper presents the main issues connected with managing cities and the problems due to different factors when compared with managing business organizations. First the differences between a city and a business organization were presented. Next the processes in both types of organization were presented. Based on the processes it was possible to present the KPIs which measure and represent the processes. There is also an important difference in the type of KPI used in a business organization and in cities. Because of this the authors proposed to create a dedicated model which will help in adapting to the needs of cities. In the last section the model was verified.

The proposed model organizes the KPIs by adding two extra levels in the structure. It helps cities to better manage the KPIs (those which are not used) and makes it possible to easily implement the system in other cities. The model was verified based on the case representing the idea of sustainable development in Warsaw.

Now it is necessary to implement the proposed changes into the software (IOC) and verify it based on the bigger amount of data.

The proposed solution can be applied for cities in which city management models are used. Then, the design process is a top-down one as described in the paper. In the absence of these models it is necessary to create them or use those existing in a high-level management model of other cities. Then city models become a specific kind of component used in the design process.

Because of this it seems expedient to modify the design processes of indicators in the IOC, a departure from the typical indicators of an organization, and the introduction of those which respond to city management processes set out in the models of city processes. This approach can be applied both at te level of tools supporting the IOC such as the Business Modeler or the Advanced version or also directly in the IOC. Then, the system designer has the ability to provide an ongoing relationship between indicators, their aggregation and the indicators necessary for the evaluation of the city processes included in the models of city processes.

It seems also to be necessary for the design process to be supported by metaphors of processes and their indicators contained in the libraries both of tools supporting the design process as well as of the IOC. Then, due to the low level of the maturity of city processes it will be possible to acquire those indicators from the libraries and directly introduce them for use in the evaluation of the processes of cities models.

References

1. A department of Economic and Social Affairs: World urbanization prospects the revision. United Nations (2014)
2. Weihrich, H.: Management – A Global Perspective. Tata McGraw-Hill, New York (2005)
3. Britannica Encyclopedia [online] (2005). http://www.britannica.com/EBchecked/topic/86277/business-organization
4. Dictionary Meriam: Webster Dictionary [online] (2016)
5. Portugali, J.: Self-organization and the city. Springer, Heidelberg (2000)
6. Maceko, M.: Specyfika Zarządzania miastem – kilka uwag. Organizacja i Zarządzanie, Silesia (2014)
7. Bettencourt, L.: The uses of big data in cities, based. In: Flood, J. (ed.) The Fires: How a Computer Formula, Big Ideas, and the Best of Intentions Burned Down New York City and Determined the Future of Cities. Penguin, Westminster (2011)
8. Bryx, M.: Innowacje w zarządzaniu miastami w Polsce. Szkoła Główna Handlowa Oficyna Wydawnicza (2013)
9. Catsella, L.: Smart cities: aspects to consider for building a model from a city government point of view. Strateg. Manage. Q. 2(3 & 4), 01–22 (2014)

10. Lombardi, P., Giordano, S., Caragliu A., Del Bo, C., Deakin, M., Nijkamp, P., Kourtit, K.: An advanced Triple-Helix network model for smart cities performance. In: Ercoskun, O.Y. (ed.) Green and Ecological Technologies for Urban Planning: Creating Smart Cities. pp. 59–72. IGI Global, Hershey (2012)
11. Government Research.: Smart Cities Model [online] (2012). www.smart-cities.eu/model.html. Accessed 2015
12. Nutter, M., Crowley, C.: How 12 cities are charting a course to being truly "smart" [online] (2015). http://www.greenbiz.com/article/12-smart-cities-Vancouver-San-Francisco-USDN-Houston
13. Madera, C.: The Future of Cities in the Internet Era. Next City, Philadelphia (2010)
14. Portal Klipfolio.: KPI in management [online] (2015). http://www.klipfolio.com/resources/kpi-examples
15. Reh, J.: Key Performance Indicators (KPI), how an organization defines and measures progress toward its goals [online] (2015). http://management.about.com/cs/generalmanagement/a/keyperfindic.htm
16. United Nations Conference on Environment & Development Agenda 21, Chap. 40 Information for decision-making (2015). http://www.un.org/earthwatch/about/docs/a21ch40.htm

Implementation of Business Processes in Smart Cities Technology

Cezary Orłowski[1(✉)], Artur Ziółkowski[1], Aleksander Orłowski[2], Paweł Kapłański[2], Tomasz Sitek[3], and Witold Pokrzywnicki[3]

[1] WSB University in Gdańsk, Gdańsk, Poland
{corlowski,aziolkowski}@wsb.gda.pl
[2] Gdansk University of Technology, Gdańsk, Poland
{aleksander.orlowski,pawel.kaplanski}@zie.pg.gda.pl
[3] Staples Advantage Poland SP. Z O.O., Gdańsk, Poland
tomasz.sitek@staples.com, witold@urtico.com

Abstract. The goal of the paper is to present the results of studies concerning the development of a method of implementation of business processes in Smart Cities systems. The method has been developed during studies carried out within the building of a Smart Cities system for Gdańsk, and is based on basic development project management mechanisms (drawing from best practices, and in particular from the RUP methodology) and business-oriented development principles, where the role of business process modeling is crucial for the implementation of functionalities of IT systems.

Keywords: Smart cites · Business process modelling · Decision support systems · Knowledge management

1 Introduction

The progress in smart information technology implies growing market interest in systems conforming to the so-called smart cities concept. Large urban agglomerations such as London, Singapore or Rio de Janeiro use IT systems to manage key processes in the city, such as road traffic management, decision-making for strategic projects or ensuring safety for residents.

Such an approach allows an urban agglomeration to be treated as an organization, in which a number of business processes occur, and these processes require the support of IT systems. A business process is usually defined as a set of operations which lead to the accomplishment of a specific goal or fulfilment of a specific business need [2]. The processes which have strategic significance for a city usually come from legislation, which provides the legal foundation for the processes carried out by a city. Among such processes, particular attention (and consequently special IT support) is attached to processes crucial for the safety and wellbeing of residents. Typically, certain procedures and guidelines which require immediate measures to be taken and the simultaneous absorbing of resources subordinate to key public order enforcement agencies such as police, fire brigade or healthcare are connected with such processes.

© Springer-Verlag GmbH Germany 2016
N.T. Nguyen et al. (Eds.): TCCI XXV, LNCS 9990, pp. 15–28, 2016.
DOI: 10.1007/978-3-662-53580-6_2

Key IT system providers strive to deliver software ("smart cities" systems) which supports the management of certain processes in urban agglomerations and their subsidiary units, such as crisis management centres. It should be noted that a crisis management centre is not an ad hoc organization, but an organization which constantly monitors certain (critical) statuses. This organization's activity intensifies when critical values defined for KPIs (Key Performance Indicators) [1] assigned to specific processes are exceeded. Also, "smart cities" systems are designed not only to support decision-making in a critical situation, but also to constantly support the control and monitoring of certain aspects of life in urban agglomerations. As an example of a status which may be monitored and which requires a response by a crisis management centre we may include the concentration level of pollutants in the air, the water level in a river, the noise level for a given location, etc.

Hence this paper aims to show how the modeling of processes is relevant for their subsequent implementation in the form of specific functionalities of smart cities systems. To that end, the current methodology knowledge supporting aspects of business modeling and architecture modeling for the purpose of IT systems has been presented. Next, the recommended model of steps leading to the implementation in smart cities systems of specific business processes carried out by the city council has been presented.

Such an approach allows the implementation and deployment of smart cities systems to be perceived from the perspective of an IT project [5], in which the implementation of system functionality is preceded by a number of activities in the area of business analysis and modeling [6]. This in turn allows a city (city council) to be treated as the client of an IT project, who has specific expectations (business needs) which should be satisfied by the functionalities of the developed smart cities system.

2   Business Modeling Issues in the Development of IT Systems

Business modeling [4, 6] is one of the main disciplines of software engineering, and is particularly emphasized in such development approaches as Rational Unified Process (RUP), but is also relevant for currently popular agile approaches [12, 13] (Fig. 1).

The disciplines in the area of business modeling (RUP includes process modeling, requirement management and architecture modeling in these disciplines) serve to analyze in detail the business needs, and consequently the business processes occurring in the organization of the client of the IT project [7], i.e. the recipient of the IT system. Business modeling as a discipline refers first of all to increasing the abstraction level of a developed system through, inter alia, the graphical presentation of processes (process modeling) and also the graphical presentation of system functionalities (use-cases diagrams). Thanks to business modeling the communication between the client and the development team is improved. Business analysts, who through their contacts with the client visualize the processes and IT system's functionalities required for subsequent implementation, are usually in charge of the business modeling.

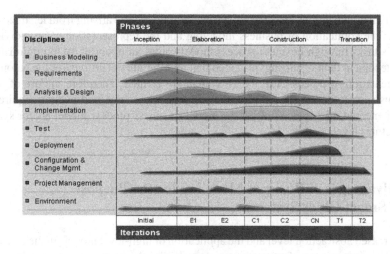

Fig. 1. Scope of business process modelling in RUP

3   Business Modeling in the Development of Smart Cities Systems

As noted in the introduction, each implementation of smart cities systems is a particular case of an IT project, which means that the disciplines connected with business modeling (business process modeling) are applicable also in the development and implementation of such systems. The application of business modeling for the purpose of smart cities systems has two main advantages.

Firstly, it allows better understanding of the functioning of the client's organization and their business needs, so that the developed/implemented IT system is properly tailored to the business processes occurring at the client. In the case of smart cities systems such an approach seems especially important. The business processes of a client like the city council are very often connected with the safety of a significant number of residents, and ensuring appropriate living conditions and standard of living. The better these processes are understood, the more likely it is that they will be properly supported by the developed smart cities system.

The second advantage of business modeling is the improved collaboration between the client and the development team (the team which develops/implements) through creating a joint platform of communication based on models and the visualization of individual items (functionalities) of the system to be built.

Another advantage, which is more and more often expected from the use of business modeling for the purpose of the implementation of IT systems is the opportunity to make the model independent from the implementation platform. This is particularly useful in the current circumstances where technical debt (a phenomenon consisting of the lack of opportunity to develop or adapt the software after some period of operation) is becoming a serious threat. Technical debt also means the risk of being unable to develop or adjust the software to the changing environment (for instance, the integration of systems in the case of two organizations merging, or system modification(s) arising from legal require-ments). The application of business modeling and the representation of an individual system's tasks through business process models gives an opportunity to minimize the risk of technical debt occurring, provided that appropriate conditions for the automatic implementation of processes in the IT system are created. Therefore the use of high-level business modeling should lead to a situation where any modification required to be implemented takes place only at the model (business process model and/or system architecture) level.

4   Advantages of Business Modeling for the Purpose of Smart Cities Systems

The increased abstraction level and the application of high-level models in the construc-tion and implementation of smart cities systems brings about certain problems, however. As noted in the introduction, one of the main problems arising from the implementation of smart cities systems in cities is the need to tailor such a system to a specific city in each case. This is due to the fact that each city is governed by individual specific formal and legal principles. In addition, each city has a number of imposed guidelines, which govern its operation in a crisis situation (these are local procedures in the case of an accident being a threat to residents).

Due to such situations, when developing and implementing a smart cities system for a particular city, the providers have to seriously take into account that it is necessary to make adjustments with respect to specific procedures, guidelines, legal acts and other instruments of a formal and legal nature. Another factor, which makes it necessary to tailor a smart cities system to the city (client) may be the specific microclimate or geographic location, which makes it necessary to modify the system's functionalities depending on the location of a particular city. Therefore, in order to avoid implementing functionalities dependent on various factors in each individual case, smart cities system providers should strive to apply appropriate components, developed, if possible, at a higher abstraction level than the source code, i.e. at the business process model level. Hence the role of business modeling leading to the presentation of individual guidelines or procedures valid in a city in the form of process models, which may be later imple-mented in smart cities systems, becomes reasonable.

The above consideration allows the conclusion that in the application of business modeling for the purpose of smart cities systems, reusable components should tend to be provided. Such components, due to their relative independence from the system proper, should allow unrestricted implementation based on procedures and guidelines delivered or held by the client (the city). Therefore in the next section particular attention is attached to a procedure model leading to the implementation of business processes defined in formal and legal documents held by the city council (Fig. 2).

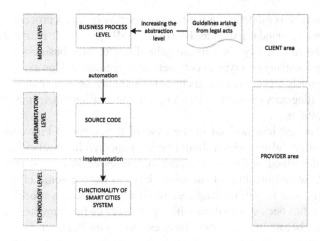

Fig. 2. Role of business modelling in the implementation of smart cities systems

5   Model of the Implementation of Business Processes in Smart Cities Systems

In connection with the above consideration it should be said that the implementation of business processes in smart cities systems should be preceded by some business modeling activities. It seems appropriate for software providers, when implementing smart cities systems, to take into account a number of key aspects mentioned above, such as the necessity to increase the system's abstraction level through the visualization of processes and functionalities. Hence the main mechanism leading to the realization of business modeling principles [11] is the defining by the client of the needs which have to be reflected (satisfied) in the smart cities system (Fig. 3).

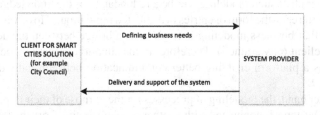

Fig. 3. Collaboration loop for smart cities solutions

The client's business need may be perceived both as a specific service provided by a smart cities system and as a specific task, which may be supported by the system. However, in order to implement such a service it is necessary to model it properly, and the best option here is modeling in the form of specific business processes with use of a certain modeling standard such as BPMN or UML [3, 14]. It is true that business processes in the case of city councils are quite different from business processes for a manufacturing company or a typical commercial organization. However, a city's activities have obviously a process-related character and may, or even should, be presented in the form of diagrams (process models) in order to facilitate their implementation in smart cities systems.

A particular problem noticed in the case of a number of city councils is that procedures and guidelines which should be implemented in smart cities systems are usually only described in the documentation (expressed in natural language). This means that those responsible for modeling have to devote some effort in order to convert descriptions in natural language into business process models. Only when this conversion has been effected, can these processes be implemented in smart cities systems. Hence before implementing business processes in smart cities systems it is necessary to analyze in detail (such analysis is mainly carried out by analysts in collaboration with city council employees) relevant procedures and guidelines in order to model them as process maps.

5.1 Issues Connected with the Implementation of Procedures and Guidelines in Smart Cities Systems

The main problem potentially encountered during business modeling for the purpose of smart cities systems is the non-typical character of business processes. In city management the processes which should be implemented in a smart cities system are different from typical development processes. In city management, the activities (processes) connected for example with responding to an exceeded critical threshold for the air or water, consist of taking the appropriate measures directly with regard to guidelines in relevant legal acts (which is shown in the corresponding diagram above). Hence the question may be asked whether business process modeling may be applicable at all for an organization like a city council.

It might seem that such modeling will be a redundancy of the knowledge which has already been entered in the form of entries in dedicated regulations. However, one should keep in mind that business modeling is to serve as a bridge between the development team and the client (city council). Therefore in this situation process modeling should be regarded as a platform enabling better communication between individual project stakeholders [9].

On the other hand, the modeling of processes for the purpose of their implementation in smart cities systems is connected with another problem arising from the formalization of such processes. If business modeling is treated as an important factor for the development of reusable components, it becomes necessary to apply certain standards for the modeling of such processes. BPMN may be mentioned as one among the most commonly applied standards today for modeling such processes; however, although users are increasingly familiar with this standard in provider organizations, this is not always the case in organizations on the part of the client.

However, the necessity to model business processes for the purpose of a city council seems to be inevitable and falls into the category of typical "business-oriented software development" principles [6], which are not only emphasized in individual IT project management methodologies, but are slowly becoming a standard in the realization of development processes. In response to these issues, a procedure model connected with the implementation of business processes for the purpose of cities which use (buy/ implement) smart cities systems has been presented below.

5.2   General Model of the Implementation of Business Processes in Smart Cities Systems

As noted above, the main problem connected with business process modeling for the purpose of smart cities systems is the necessity to carry out a multi-aspect (multi-thread) business analysis. On one hand, those responsible for modeling deal with the

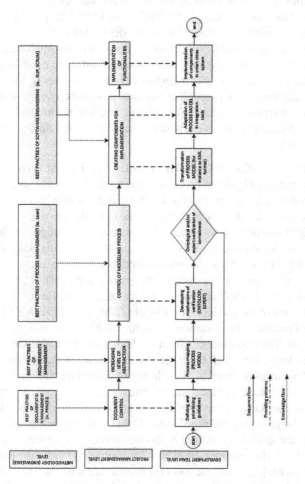

Fig. 4. General model of the business processes in smart cities systems

documentation (legal acts, standards, regulations), which must be reflected in the system. On the other hand, good software engineering practices make it necessary to present these aspects contained in the legal acts at a sufficiently high level of abstraction. Thus it means that when intending to model business processes and implement them for the purpose of smart cities systems, at least three organizational levels should be taken into account, which influence the realization of modeling and implementation activities.

The first, and the highest level, is connected with best practices of procedure — both for documentation management and best development practices. This level may be called the knowledge level and it is designed to ensure methodological best practices of procedure. However, it should be kept in mind that the implementation of smart cities systems is a typical IT project, and thus a number of activities have to be carried out in accordance with project principles (managing, control, monitoring progress) at the operation level (project management). And finally, at the lowest level, individual activities are carried out which enable the creation of process models and their implementation in smart cities systems. Such an approach is in accordance with hierarchical project management represented for example by PMBoK or PRINCE2 [9, 10].

These three key levels are reflected in the model shown in Fig. 4.

As noted above, the procedure model for the purpose of the implementation of business processes in smart cities systems takes into account three main levels, which are interrelated. Each of these levels is a certain organizational level. Such an approach makes it easier for the persons modeling business processes to analyse the knowledge sources supporting specific activities (knowledge level) in the modeling area, and also indicates the activities (project level), which should be effected during the development of smart cities systems.

- METHODOLOGY (KNOWLEDGE) LEVEL is the highest layer of the model of the implementation of business processes; this layer is to provide appropriate knowledge required in order to carry out modeling and implementation processes. Hence the presence of this layer in the model gives the modeling persons and implementing persons insight into necessary sources of substantive knowledge used in the management of development projects. This layer is aimed at ensuring, inter alia, best practices applied in IT projects, with a recommendation to make use of approaches such as PRINCE2 (for instance for documentation management), RUP (increasing the abstraction level), LEAN (improving what was already created).
- The second level of the model is the PROJECT MANAGEMENT LEVEL, i.e. the realization of particular organizational activities aimed at the delivery of the system's functionalities. The best practices from the higher level provide knowledge for this management level, for which the project manager and/or the steering committee is in charge. This level is aimed at the delivery of procedure patterns to be applied by the development team creating smart cities solutions. The patterns, which are supplied from the management level, have been divided into the following: document control (i.e. the gathering and processing of legal acts, regulations and other sources, from which the guidelines are derived), the provision of patterns to increase the abstraction level (for instance BPMN, UML notations) and also the provision of patterns for control, for example tools (in accordance with the TOOLS & TECHNIQUES principle) for the verification of developed models. Finally, at the management

level attention should be paid to design patterns (in RUP these is called capability patterns), i.e. a recommendation to apply reusable components, which later may be implemented autonomously in subsequent systems. Thus, the last set of management activities is the realization of the implementation processes, reflecting the development level.

- DEVELOPMENT TEAM LEVEL, i.e. the model level dedicated to the development team in order to deliver some procedures of operation (autonomously developed by the authors and discussed in detail in a further section). At this level, firstly one should refer to the interpretation of guidelines, which are to be modeled in the form of a business process, which subsequently should be verified (the team has developed a dedicated tool for verifying the conformity of the process map/model with the documentation on the basis of ontologies). Next, when the process model is in accordance with the guidelines (documentation), it may be converted into an interpretable form by means of dedicated tools applied for the development of smart cities systems. Due to the high degree of complexity of this level, it has been further extended as a detailed model.

5.3   Detailed Model of the Implementation of Business Processes

The developed general model indicates the main layers which are required in order to carry out correctly the business modeling for the purpose of the implementation of smart cities systems. In addition, the modeling persons (business analysts, city council employees) should take into account a number of additional activities (recommended by the authors of this paper) enabling the correct modeling of procedures and guidelines, which may be later implemented in smart cities systems. Hence the detailed model considers the actors (participants) in individual activities carried out for the purpose of the implementation of business processes in smart cities systems, presents the sequence of activities, and also shows the processes which are of key significance for the modelling, occurring within such areas as defining business needs, process modeling, the verification and control of correctness, and implementation (Fig. 5).

The detailed concept of the model also contains, apart from the main areas connected with process modeling and implementation, responsibility for individual activities leading to the implementation of processes in smart cities systems. Hence a relevant city council employee is responsible for providing documents with procedures and guidelines necessary for the subsequent modeling of business processes, which will be implemented in the smart cities system. It should be noted that although the guidelines come from legal acts (like charters, regulations) held by the city council, it is the particular employee who defines their business need (i.e. which items the system should contain so that employees are able to fulfil their duties arising from the legal acts).

When the guidelines have been prepared, the business process modeling stage takes place. A process analyst (who may be indicated by the city council or may come from the provider's organization) through consultation/collaboration with the city council employee begins the business process modeling. In a mature organization, processes would be modeled by department employees, while in a less mature organization competent consultants should also be involved. In the process modeling area indicated

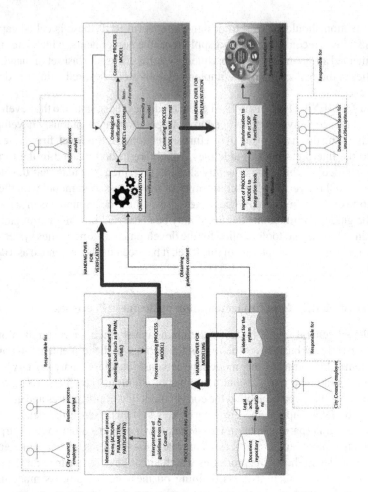

Fig. 5. Detailed model of the business processes in smart cities systems

in the model, in the beginning, the relevant guidelines in the documentation provided by city council employees are interpreted. The interpretation is a process strictly connected with activities such as the analysis of the content of an act/ordinance, on the basis of which the subsequent system functionalities are to be created. Next, all the elements within the process, namely the entities (process participants), the actions taken by the entities (process activities) and the conditions (process parameters) such as permissible pollution concentration levels, etc. are identified (through an analysis of the text). When all the elements within the process have been determined, the appropriate modeling standard should be selected. A certain standard is necessary in order to avoid disputes as to workflow within the process (hence the present standards such as BPMN or UML are recommended). After having selected the modeling standard (notation) and modeling tool (in accordance with best project management practices [9]), process mapping can proceed, i.e. a graphical presentation of the process within the tool.

Another area in which the tasks required for the implementation of business processes in smart cities systems are carried out, is the verification of the previously developed business process model. In this paper a verification based on ontologies [8] (based on the experience of the authors) is recommended. In the studies carried out by the authors, mechanisms have been developed of automatic verification of business process models on the basis of linguistic descriptions based on the dedicated ONTO-TRANSTOOL tool. This tool, on the basis of a description in natural language, identifies all entities, i.e. process elements contained in legal acts/ordinances, and then it is checked whether the processes modeled and generated by ontologies are convergent with those modeled by the analyst on the basis of contacts with a representative of the city council. This control is carried out expertly by the analyst. Should any discrepancy be found, the process will be remodeled.

After having verified the process model the export file should be prepared (based for example on XML standard) for the subsequent importing thereof by dedicated tools integrating processes with smart cities systems, such as IOC. Therefore the last area of the procedure model is connected with the integration and implementation of already modeled business processes in the smart cities system. By means of a dedicated tool one can import the process model written in xml format, and then send the model imported from the integration tool to the person in charge of the implementation of business models in the smart cities system (in the case of IOC it is the appropriate INTEGRATOR and BUSINESS MONITOR). Only after having imported in the BUSINESS MONITOR is it possible to generate the desired functionalities corresponding to the needs from the BUSINESS NEEDS AREA, such as Key Performance Indicators (KPIs) or Standard Operating Procedures (SOP).

6 Verification of the Model of the Implementation of Business Processes in IBM IOC

In this section an example concerning a threat arising from exceeded PM10 concentration in the air for the Tri-City agglomeration, the city of Gdańsk, Poland has been used, on the basis of the guidelines provided by Gdańsk City Council, as an example to illustrate the opportunities for the implementation of business processes in smart cities systems. In connection with the prior considerations it should be noted that in the case of urban agglomerations most activities showing symptoms of a business process arise from guidelines and procedures being formal and legal constraints for the decision makers and users of smart cities systems. This section aims to present how to automate the implementation of guidelines, which are represented in the form of a business process model, within smart cities systems.

EXAMPLE. During the work carried out for the purpose of the implementation of such a system for the city of Gdańsk, a procedure was used describing certain measures to be taken in a situation where one of two thresholds for PM10 concentration is exceeded. When the lower threshold (in the analyzed case the threshold is 50 $\mu g/m^3$) is exceeded, it means the occurrence of the first threat level, and the action to be taken consists first of all of notifying the appropriate units. When the second threshold (200 µg/m³) is exceeded, more units are notified and certain entities (such as the fire brigade) are alerted. The entities to be notified and alerted are specified in the relevant regulations.

As can be seen in the description above, the guidelines (procedure of conduct) include several main elements, which allow them to be treated as business processes. Each procedure (guidelines) is composed of the following elements:

- actions (measures to be taken by the smart cities system operator when a critical situation has occurred),
- parameters (critical values which stipulate defined measures to be taken),
- process participants (individual entities taking measures or notified after a critical situation x has occurred).

The persons who have been notified by the crisis management centre (i.e. the entity which manages the information flow) that a certain event has occurred, operate in accordance with their charter-based duties and in-house procedures. Within the studies carried out on the basis of the IBM IOC platform, the implementation of such processes may take place only through such tools which may be integrated with the IOC platform.

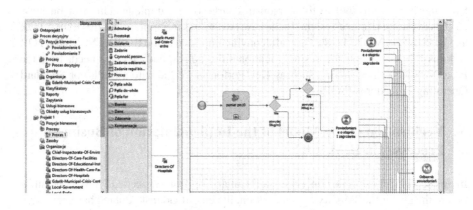

...

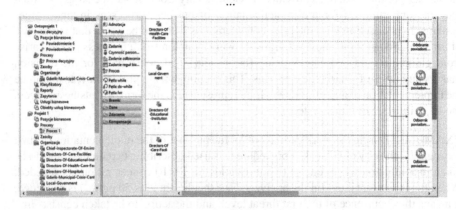

Fig. 6. Model of the process in WEBSPHERE MODELER

The IBM Websphere Modeler (Advanced version) is a tool dedicated for such purposes. During the studies the guidelines concerning an exceeded PM10 concentration level were modeled within this tool, in accordance with the process definition contained in the documentation provided by the city council. The model of the above mentioned procedure is shown in Fig. 6.

Next, the model developed within the tool is confronted with the process model generated by the ONTOTRANSTOOL tool in order to verify correctness. In the case of minor differences the modeling person could correct the already modeled process, if necessary. At this stage of the studies, the checking of the conformity of the process modeled by the analyst with the one generated automatically by the ontologies takes place manually. The differences which have been noticed during the verification of the developed procedure model for the implementation of processes could be for example attributed to the fact that in the documentation a group of entities may be indicated as a single participant in the project, while the tool based on the ontologies [8] treats these entities as independent items (for example the notified group comprises directors of schools and kindergartens). From the point of view of the person modeling the business process, this may be a single track of the process, and from the point of view of the ontology, directors of schools and kindergartens are separate items, and thus separate tracks in the process. This does not influence the subsequent implementation of functionalities (both entities will be notified about a threat as defined in the relevant procedure), however, the suggestion to specify accurately the recipients of the information may serve as a hint for persons who define procedures in the city council.

7   Summary and Recommendations

The goal of the above consideration was to indicate the role of business modeling in the development of smart cities systems. The developed procedure model for the implementation of business processes indicates the key areas for managing documents with guidelines. On the basis of the guidelines it is recommended to develop a process model which will reflect the procedure of conduct (for example in a crisis situation) contained in the documentation. Next, it is recommended to verify the developed process model (the authors have applied a dedicated tool based on domain ontologies) for correctness with the formal and legal provisions in the documentation, and only afterwards to proceed with the actual implementation.

Such a solution enables the correct implementation of business processes, and consequently, the system functionalities would be better tailored to the users' (clients') expectations. Also, when we use the process model, any modifications are made to the model (and not in the code) and therefore the modifications are better understood by both the client and the development team.

The application of business modeling for the purpose of smart cities may be a response to the technical debt risk and may serve as a certain platform providing protection against a shift in technology. When we use a modeling standard (for example BPMN or UML) and later ensure a mechanism for the translation of the model in a specific standard into integration tools (which is recommended by the authors), it is very likely that we will be protected against technical debt risk. Such an approach also means an opportunity to make better adjustments to any changes in the law and provides more opportunities to the software provider to reuse such previously created components (for example when in another city the procedure is similar, but criteria values are different). Further work on the developed model should be directed towards full automation of the implementation of business processes, i.e. the transformation of the already modeled process to the source code. The bridge solutions applied by the authors require at least two additional tools — an integrator and a model management system. Full automation would allow independence from other technologies; this would speed up the implementation processes of the functionalities, and in the future might provide an opportunity for the autonomous management of the system's functionalities by the users (for example city council employees, who by modeling the process themselves ensure its realization in the smart cities system).

Is also seems reasonable to strive to unify modeling standards, and, in the case of platforms such as the IBM IOC platform applied by the authors, a greater openness to process models created in other notations than those contained in the related tools would be welcome. Thanks to that there is a chance that the recipients (cities) that have already modeled their processes in the BPMN standard will not have to transfer them to the modeler's own notation.

References

1. Bhowmick, A.: IBM Intelligent Operations Center for Smarter Cities Administration Guide 5. Event flow diagnostic and validation tool for IBM WebSphere Business Monitor. International Business Machines Corporation (2009)
2. Bitkowska, A.: Zarządzanie procesami biznesowymi w przedsiębiorstwie, Vizja Press & IT (2009)
3. Drejewicz, S.: Zrozumieć BPMN. Helion (2012)
4. Górski, J.: Inżynieria oprogramowania w projektowaniu informatycznym, Mikom (1999)
5. Kowalczuk, Z., Orłowski, C.: Design of knowledge-based systems in environmental engineering. Cybern. Syst. Int. J. 35(5–6), 487–498 (2004)
6. Kruchten, P.: The Rational Unified Process: An Introduction. Addison-Wesley, Boston (2004)
7. Orłowski, C., Kowalczuk, Z.: Modelowanie Procesów Zarządzania Technologiami Informatycznymi. Pomorskie Wydawnictwo Naukowo-Techniczne PWNT. (Automatyka Informatyka). Książka profesorska (2012). ISBN: 978-83-926806-4-2
8. Orłowski, C., Ziółkowski, A., Czarnecki, A.: Validation of an agent and ontology-based information technology assessment system. Cybernetics and Systems 41(1), 62–74 (2010). 5 rys. - Bibliogr. 7 poz. – ISSN: 0196-9722
9. PMI.: A guide to the project management body of knowledge, 5th edn. USA (2013)
10. PRINCE2®. Managing Successful Projects with PRINCE2® (2009)
11. Sacha, K.: Inżynieria oprogramowania. PWN (2010)
12. Schwaber, K., Sutherland, J.: SCRUM Guide (2013)
13. Schwaber, K.: Sprawne zarządzanie projektami metodą Scrum (2005)
14. Wrycza, S.: Język UML w modelowaniu systemów informatycznych. Helion (2008)

Designing Aggregate KPIs as a Method of Implementing Decision-Making Processes in the Management of Smart Cities

Cezary Orłowski[1(✉)], Artur Ziółkowski[1], Aleksander Orłowski[2],
Paweł Kapłański[2], Tomasz Sitek[3], and Witold Pokrzywnicki[3]

[1] WSB University in Gdańsk, Gdańsk, Poland
{corlowski, aziolkowski}@wsb.gda.pl
[2] Gdansk University of Technology, Gdańsk, Poland
{aleksander.orlowski, pawel.kaplanski}@zie.pg.gda.pl
[3] Staples Advantage Poland SP. Z O.O, Gdańsk, Poland
tomasz.sitek@staples.com, witold@urtico.com

Abstract. The aim of the paper is to present a concept of measuring the performance of city management processes by use of a concept of aggregate KPIs. In the management of organizations and, as a consequence of the use of a common design framework also in the management of cities, silo KPIs are commonly used to show the statuses of the processes of organizations/cities. Thus the question arises as to what extent aggregate KPIs, as proposed in the paper, can be used in the management processes of smart cities in place of the silo ones typical for organizations. The work is divided into four main parts. The first presents the problems of managing smart cities to introduce the reader to the problems of measuring processes and the need for aggregated measurements. The second section discusses KPIs and their place and role in management processes. The third part contains a description of the model of aggregate KPIs to support measurements of the status of city processes. In the fourth section the developed model is verified, demonstrating its applicability for city management processes. The summary includes a recommendation for the use of aggregate KPIs in the city.

Keywords: Smart cites · Knowledge base · Knowledge management · Fuzzy logic · Process modeling · Decision support

1 Introduction

Modern cities now occupy just 2 percent of the earth's surface and are home to as many as 50 percent of the world's population. It is expected that by 2050 already 70 percent of mankind will live in cities. According to data from the Central Statistical Office in Poland, already today the number is about 60 percent [4]. In addition, the number of people migrating from an urban to a metropolitan area is steadily increasing. All this results in the management of large cities becoming a challenge of modern civilization. It is dependent on experience, competence and above all available resources within the

© Springer-Verlag GmbH Germany 2016
N.T. Nguyen et al. (Eds.): TCCI XXV, LNCS 9990, pp. 29–42, 2016.
DOI: 10.1007/978-3-662-53580-6_3

agglomeration. In many cities, decision-makers are widely supported by information technology in the analysis of their decision-making processes [10].

Although in the context of different conditions of the functioning of cities and their development many approaches and concepts of city management can be seen, they are increasingly treated as manageable in the context of the application of intelligent processes to manage their operation. A Smart (or Smarter) City can be described as an idea the foundation of which is the implementation of these processes. There is no single agreed definition of a Smart City. It should be assumed that it is rather a vision based on two pillars - best management practices and opportunities to support city processes with broadly defined information technology [6].

The implementation of IT in any large organization is an undertaking for which at least three limitations should be determined: the scope of work, the schedule of the project and its budget. From the point of view of such a large and complicated structure as a city, what is particularly important is the scope of work that is desired by the recipient - the functionality of the city system. The management of such a structure must first and foremost require the functionalities to be systematized and described and the most important ones to be selected. The identification of key functionalities for system implementation may be problematic in the case of these processes, the implementation of which involves several entities of the city. Then, the implementation of IT solutions is not synchronised by particular entities at different levels of local government, leading to a high degree of fragmentation of the processes and systems of the city.

The authors set themselves the goal of designing and implementing the system in Gdansk. It was assumed that the aim will be to design the components of the IBM IOC (Intelligent Operating Centre) project framework, which can be used in the case of other cities [3]. This paper focuses on the design of one of the components - KPIs as a measure of the status of the city's decision-making processes.

2   Examples of City Decision-Making Processes Supported by Silo KPIs

KPIs (Key Performance Indicators) are defined as measures of processes regarding the achievement of an organization's objectives. This concept appears most often in the context of financial data and acts as a signal to decision-makers about the status of work processes, their cost or quality. They may take the form of managerial control tools and should create conditions to make decisions without burdening decision-makers with a detailed analysis of the source data. In practice, many different KPIs are defined for business use: financial indicators (e.g. the margin calculated per customer, the sales value calculated per employee), in the customer service area (% of overdue deliveries to customers), quality of service (number of complaints) and many more [8]. It should be noted that a KPI is a function of any input data which is relevant in terms of the decisions taken. What is important in the context is the scale predetermined for each indicator identifying critical values (the exceeding of which should be treated as an emergency situation) [7].

KPIs are an element of most information systems, which assist the management of the organization at various levels (operational to strategic). They are implemented in

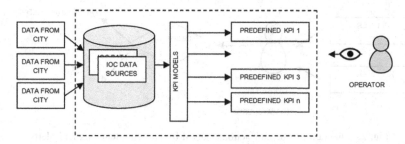

Fig. 1. Systems of Smart Cities - sequence of KPI design processes

systems such as ERP (Enterprise Resource Planning) or CRM (Customer Relationship Management). They are also used in the systems of Smart Cities. In such systems, the structure of KPIs covers the following sequence of processes:

- the identification of external data sources feeding the internal database of the system of Smart Cities (including - establishing timetables for their acquisition),
- the construction of KPI models that define relationships between different input data (one KPI model can be used to build a family of a number of KPIs),
- the development and definition of individual KPIs, which will be available on the desktop of the system operator (including setting the scale presented by the system through a range of several colors).

The above-described sequence of processes leads to the running of cyclic data flow in the systems of Smart Cities. This sequence is illustrated in Fig. 1. The source data is obtained from external sources which supply the internal database of Smart Cities (process 1). Figure 1 shows a sample database implementation for the IBM IOC system. This database itself is a source of data for KPI models (process 2). KPI models represent logically (thematically) grouped data for the needs of specific KPIs (process 3). Each defined KPI becomes an object accessible to the operator from his desktop. The operator, however - from the point of view of decision-making processes, becomes a "KPI watcher", and only then - a decision-maker [1].

The above-described indicator design process is an example of the design of so-called silo KPIs. They work well for the implementation of procedures that do not require immediate action. Much of the data may in fact be obtained from the city with a delay with respect to measurements. Many measurements are not made in a continuous cycle, so the operator does not receive them in the system in real time (e.g. meteorological measurements from sensors). Such an approach is acceptable only for a part of the city procedures. In the event of an incident posing a threat to human health and life, the effectiveness of the relevant services coordinated by the operator directly depends on the speed of obtaining data. Then support for decisions is understood as the following sequence of processes:

- immediately providing the operator with information about the existence of the threat (by launching the KPIs),

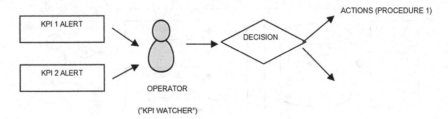

Fig. 2. Diagram of operator actions caused by two simultaneous KPI alerts

- supporting the operator in the rapid notification of the relevant services (in accordance with the obligation that in a given crisis situation is imposed on the city by a proper act of law),
- presenting options and suggestions for decision-making processes - in particular, the allocation of resources such as ambulances, the fire brigade, etc.

In the cases described above, a KPI is a source of information (knowledge) related to a specific single critical situation, which is easy to model. The system designer in the creation of a model of a silo KPI needs specific (key for the process) data from external sources, an established method of data processing and the determination of an item on his desktop along with the presentation conditions of this element (the scale and limit values). However, if we assume that in an organization like a city every day many emergencies occur simultaneously, then it is a decision-making problem for the operator when he receives e.g. two alerts (emergency notifications). He must then run two parallel procedures, with the result that at the system level, there is a need to integrate data from external sources which are unrelated. In the absence of such integration, decision support must be implemented through many different KPIs. An example diagram of the actions of an operator caused by two simultaneous KPI alerts is shown in Fig. 2.

Figure 2 presents a situation in which the operator should make a decision (the allocation of resources available in a limited amount) for two emergencies signaled by two different KPIs. These two data sources, disjointed for the operator, mean that he must perform an aggregation of both KPIs. Such a solution is a heavy burden for the operator and at the same time increases the uncertainty of decision-making. However, based on an analysis of the situation shown in Fig. 2 the following conclusions can be formulated regarding the impact of a KPI analysis on the decision-making processes of an organization (in particular - a city):

- a KPI, understood as a system interface element indicating a critical situation on the basis of simple arithmetic, takes the form of a silo [2]. It acts as an alert to the operator, but by itself does not support his decision. In situations where there is a real threat to human life, this solution is therefore too "slow".
- KPIs are tracked individually by the operator. A KPI performs a separate calculation of an incoming value in relation to the threshold value; similarly each indicator

Fig. 3. Silo KPIs of the IOC system for the emergency landing of an aircraft, exceeding PM10 and the use of ambulances.

is displayed as a separate object. A screen with multiple KPIs available to the operator may cease to be legible (see Fig. 3, which shows an example of a system desktop with several KPIs defined).

- KPIs can thus be identified with so-called information silos - being an accurate view, but only of a slice of the critical area of the city.

The above observations are well illustrated in Fig. 3, which is a screenshot of the IBM IOC desktop. It shows the silo KPIs as a set of elements (rectangles) regarding three different situations: the emergency landing of an airplane, exceeding PM10 and the use of ambulances. The color range is set at the construction stage of the KPIs.

The above-mentioned KPIs represent status images of information silos and do not depend on each other. They illustrate the statuses of processes, but in a detached way. However, in such an organization as a city, the city processes are closely dependent on each other, so the idea of the presented KPIs is not fully reflected in the city from an analysis of the status of city processes [9]. Therefore, the aim of this work is to present aggregate KPIs which fully, not in part, will be used in city decision-making processes. They should be used by the operator/decision maker and their design should be based on models of city processes. The model of aggregate KPIs is discussed in the next section of this paper.

3   Models of Aggregate KPIs

In the previous section silo KPIs based on individual data were defined at the implementation stage of the Smart Cities system. Currently, the concept of integrated KPIs will be presented based mainly on models of the city processes, which form a basis for creating logical chains of KPIs designed in a Business Monitor. This type of KPI is used to create dynamic structures to support decision-making processes. A generic model of these indicators was developed for the implementation of decision-making processes for any city.

3.1   Aggregate KPIs - Definition and Meaning

The silo KPIs which were presented in the previous section of this paper are subject to procedures for processing data in information silos. Unlike them, aggregate KPIs are

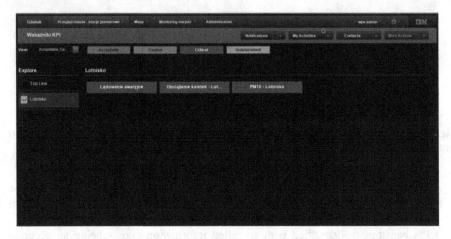

Fig. 4. Initiating and supporting KPIs as components of aggregate KPIs

structures which operate further in a specific context of decision-making - being the currently implemented decision-making processes of the city (activating specific operating procedures). The launching of aggregate KPIs is determinant to a certain, specific procedure, which constitutes a response to a detected threat (real or potential).

To further explain the concept of aggregate KPIs, it is necessary to clarify important concepts. The paper considers aggregate KPIs as a sequence of processes, which include:

Decisions - as a consequence of choices faced by the operator; they affect the processes of the city

Actions - which depend on the decisions taken and are seen as consequences of the decisions and are depicted as results of KPIs which:

- induce procedures (hereinafter referred to as initiating KPIs - KPI_I) - exceeding the critical values presented by them makes the first step in each of the procedures. Examples of KPI_I: exceeding the safe levels of airborne concentrations of air pollutants (e.g. PM10).
- provide information helpful in decision-making (called supporting KPIs - KPI_W) and are necessary in the decision-making process. In the studied cases KPI_W was identified with the number of available resources necessary to reduce or eliminate the emergency. Examples of KPI_W: the number of available ambulances in the city, the number of beds in public medical facilities.

Example initiating KPIs (emergency landing and exceeding PM10 levels) and supporting KPIs (number of ambulances) are presented in Fig. 4.

The following presents the structure of a decision-making chain showing how aggregate KPIs are constructed (Formula 1).

```
if      (Lądowanie_x0020_awaryjne>=70      and      PM10_x0020_-
_x0020_Lotnisko>=60) then 2
  else   if   (Lądowanie_x0020_awaryjne>=70   and   PM10_x0020_-
_x0020_Lotnisko>=30) then 2
  else   if   (Lądowanie_x0020_awaryjne>=70   and   PM10_x0020_-
_x0020_Lotnisko>=0) then 1
  else   if   (Lądowanie_x0020_awaryjne>=40   and   PM10_x0020_-
_x0020_Lotnisko>=60) then 1
  else   if   (Lądowanie_x0020_awaryjne>=40   and   PM10_x0020_-
_x0020_Lotnisko>=30) then 1
  else   if   (Lądowanie_x0020_awaryjne>=40   and   PM10_x0020_-
_x0020_Lotnisko>=0) then 1
  else   if   (Lądowanie_x0020_awaryjne>=0   and   PM10_x0020_-
_x0020_Lotnisko>=0) then 0
  else   if   (Lądowanie_x0020_awaryjne>=0   and   PM10_x0020_-
_x0020_Lotnisko>=1) then 0
  else 0
```

$$(1)$$

It covers both the procedures (described above) as well as taking into account the levels of risk. The condition for the recognition of a situation as dangerous and the launching of the due process in this respect is the case of only one initiating KPI (KPI_I) exceeding (e.g. the level of PM10). It is enough to receive a signal that the KPI values have been exceeded to recognize a situation as dangerous. In the above procedure (Formula 1) it becomes necessary to acquire the values of many supporting KPIs (KPI_W). The model of the relationship between various types of KPI and the procedure of conduct, which is a generalization of the processes shown in Figs. 3 and 4, is shown in Fig. 5. In this Figure, the axis representing performance during the procedure to be conducted presents initiating KPIs (at the beginning - this indicator starts a sequence of actions) and supporting KPIs (during the procedure - when making decisions).

In the case of a single threat silo KPIs still seem to be sufficient - both in the role of initiating and supporting. A silo KPI does not require the context of the processes mentioned above, it is always launched in connection with one particular decision-making process. However, it should be noted that in the case of a simultaneous occurrence of two (or more) threats the simple operational model becomes inefficient, and in fact, almost impossible to use. The operator receives two different suggestions of decisions on the allocation of resources (e.g. the mentioned ambulances) from two

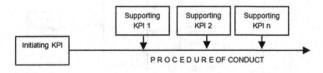

Fig. 5. Model of relationships between initiating KPIs and supporting KPIs and the procedure of conduct developed on the basis of Figs. 3 and 4

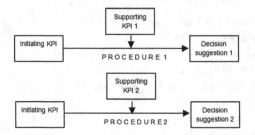

Fig. 6. Model of support for two simultaneously launched procedures by silo KPIs - two independent decision suggestions

different procedures in the system. Unfortunately, neither of the procedures takes into account the existence of the other. In extreme cases, it may be that for both threats the operator will be prompted to allocate the maximum available resources (i.e. for example sending all available ambulances to two different places). The decision support model for the two processes supported by silo KPIs is shown in Fig. 6.

The solution for the problem presented in Fig. 6 is a decision support model that uses aggregate indicators. Their essence and meaning will be described in the next section.

3.2   Model of Aggregate KPIs in Support of City Procedures

The most important premise at the basis of the structure of this model is the elimination of independent decisions, being a consequence of the applied procedures. It is assumed that procedures that the operator coordinates cannot be considered separately because there are common factors of both decisions pictured by the aggregate indicator (KPI_Z). KPI_Z will present the state of data taking account of many city processes varied as to value and size. The functioning of aggregate KPIs is shown in Fig. 7.

Hence, the models presented in Figs. 6 and 7, although based on the same initial assumptions, generate two different types of decision suggestions (two independent or one cumulative), because:

- The operator, in both cases, faces two notifications (initiating KPIs).
- The operator uses two simultaneous decision suggestions (using two supporting KPIs).

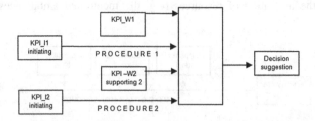

Fig. 7. The proposed city decision support model for several simultaneously launched procedures integrated through the aggregate KPI_Z

The cumulative decision is the result of the aggregation of initiating and supporting KPIs. They form the integrated KPI_Z. As previously assumed, decision support is understood here as the separation by the operator of scarce resources needed to implement the accepted procedures (Fig. 7). Any procedure which is a response to a critical situation in the city can be seen as a pair of (two) elements described as:

$$P_x[KPI\_I_{1x}, (KPI\_W_{1x}, KPI\_W_{2x}, \ldots, KPI\_W_{nx})] \tag{2}$$

where:

Px – procedure X
KPI_I_{1x} - KPI initiating procedure P_X
KPI_W_{nx} - KPI set to support procedure P_X

It should be noted that in different procedures the same KPI_W (supporting) may be used. Hence, the first step in the stages of the implementation of Smart Cities can be the preparation of a KPI_W directory. In this way, the aggregate indicator KPI_Z will be able to use an identical KPI_W for two analyzed procedures. This approach creates conditions for the allocation of the same kind of resources (e.g. both procedures require information on the number of ambulances, so for both, a decision must be taken concerning the same resources). Such a catalog of KPIs can provide a basis for the ontological description of KPIs.

The proposed aggregate indicator KPI_Z will therefore present the value (result) dependent on the status of the indicators initiating all the procedures P_x that are being initiated at a given time. The process of formalizing indicators and procedures defined in such a way creates conditions for the integration of all the threats that exist at the same time and require decisions concerning the allocation of resources. The aggregate KPI_Z is therefore a function dependent on the procedures P_1, P_2,... P_n that it integrates. Thus, the KPI_Z indicator for a particular procedure Px will adopt a value which depends on the number of resources that can be allocated to the procedure x.

$$KPI\_Z = f(P_1, P_2, \ldots, P_n) \tag{3}$$

It is important to clarify on what basis the KPI_Z indicator will adopt certain values. These values may be adopted at the stage of the modeling and implementation of each procedure with the use of a so-called KPI dependence matrix and during the updating of the priorities matrix. Both of these matrices form a knowledge base to complement the processes of modeling.

3.3   Dependence Matrices of Aggregate KPIs

KPI dependence matrices are objects that inherently contain knowledge about relationships between various indicators which a given procedure uses. Thanks to their creation, a KPI designer will directly identify connections between KPIs and city processes. This dependence becomes essential for the proper utilization of the KPI_Z indicator. Therefore, in building a dependence matrix for each procedure, it will be

Table 1. Dependence matrix KPI_I - KPI_W for a single procedure

	KPI_I_1	KPI_I_2	KPI_I_m
KPI_W_1	Z_{11}	Z_{21}	Z_{m1}
KPI_W_2	Z_{12}	Z_{22}	Z_{m2}
KPI_W_n	Z_{1n}	Z_{2n}	Z_{mn}

necessary to determine which KPI_I and which KPI_W are necessary for the proper operation of aggregate KPIs. An example KPI dependence matrix is shown in Table 1.

It is also assumed that:

$$Z_{XY} \in \{0, 1\} \tag{4}$$

where:

Z_{XY} - a relationship between procedures X and Y expressed in a binary way;

A value of 0 indicates no correlation in the given procedure between KPI_I and KPI_W indicators while a value of 1 means that such a correlation exists. A threshold of acceptability of the value of this factor is not specified.

The KPI dependence matrix described in this section is designed to link existing indicators of different procedures. The system will then be able to use that knowledge at the stage of checking to what extent two critical situations require the same kind of resources to be allocated. If it turns out that such a requirement exists, the operator will face a decision concerning the allocation of a specific number of resources to be directed to two different locations in the city. To determine which resources are of higher priority it becomes necessary to construct a priorities matrix. The method of its construction is detailed in the next section.

3.4   Priorities Matrix of Aggregate KPIs

If it turns out that there is a need to make a decision on the allocation of resources of the same type, it should be determined which part of the resources should be assigned to which procedure. Therefore it becomes necessary to define the rules for the determination of the KPI_Z indicator for each of the procedures. At the implementation stage of the procedure, the priority must be specified with regard to other procedures. Such knowledge is stored in the priorities matrix (Table 2).

At this stage it is assumed that for each procedure P_x a weight relative to all other procedures will be specified. It is a simplified assumption that certain critical situations will arbitrarily have a higher priority than others at the resource allocation phase.

Table 2. The priorities matrix for procedures running simultaneously

	P_1	P_2	P_n
P_1	0	W_{12}	W_{1n}
P_2	W_{21}	0	W_{2n}
P_n	W_{n1}	W_{n2}	0

The W_{XY} values included in the table represent the priority of procedure X with regard to procedure Y. It is also assumed that the following conditions must be met:

$$W_{XY} > 0, gdy X <> Y \tag{5}$$

$$W_{XY} = 0, gdy X \leq Y \tag{6}$$

$$W_{XY} = \frac{1}{W_{YX}} \tag{7}$$

It is also assumed that the value of W_{XY} is important and used to calculate the KPI_Z, where at the same time two premises occur:

Both procedures X and Y are launched simultaneously.

Both procedures X and Y use the same KPI_W indicators (i.e. require resources of the same type).

The value searched for an aggregate KPI_Z indicator for procedure X will then be determined as follows:

$$KPI\_Z_X(P_X) = W_{XY} * KPI\_W_n \tag{8}$$

The indicator calculated according to the Formula (8) uses both pre-defined KPIs and knowledge contained in both KPI matrices - dependence and priorities. It should be emphasized that the presented method of determining the indicators should be verified. For the purposes of this process, the occurrence in the city of two different threats was assumed and it was shown how the aggregate KPI_Z indicator can be applied in this case to support the decision maker - the operator of the Smart Cities system.

4 The Use of the Aggregate KPI Model in the Process of Designing the Smart Cities System

In order to verify the model presented in this study, we analyzed the potential situation of the simultaneous occurrence of two threats of a different nature. Despite the differences, both cases may pose a real threat to the residents of the city and both are defined in the city procedures of conduct. The following two procedures were considered:

- Procedure P_1: launched, if it is found that the concentration of hazardous substances in the air (in particular - PM10) has exceeded a level safe for residents - the appropriate KPI_I_1 is triggered when one of the measuring stations (sensors) located in the city provides a series of data regarding the concentration of PM10 higher than the established threshold; such a situation may adversely affect the well-being and health of the residents located in the area covered by the event. Recommended actions/decisions for the operator: to proactively send ambulances to the site.
- Procedure P_2: when a signal is received from airport services concerning problems with an airliner approaching for landing (KPI_I_2 launched); passenger lives in danger. Recommended actions/decisions for the operator: to proactively send ambulances to the airport (just as in procedure P_1).

The decision problem for the operator is therefore to select the adequate number of ambulances to be sent to each place. For such a simplified example the following variables will thus be specified, expressed in the IBM IOC system with KPIs:

- KPI_Z_1 and KPI_Z_2 - searched values - the number of ambulances suggested to be sent respectively for procedures P_1 and P_2,
- KPI_I_1 - the indicator initiating procedure P_1, i.e. exceeding the permissible concentration level of PM10,
- KPI_I_2 - the indicator initiating procedure P_2, i.e. information about danger to the aircraft,
- KPI_W_1 - the number of ambulances available at a given time, information acquired from the medical services. For this example, it is assumed that KPI_W_1 = 100.

As both procedures use the same KPI_W indicator, the dependence matrix Z between KPIs looks like that shown in Table 3.

Table 3. Dependence matrix Z for the test case

	KPI_I_1	KPI_I_2
KPI_W_1	1	1

For this case it is assumed that procedure P_1 has a much higher priority than procedure P_2 in terms of the resource allocation (Table 4).

Table 4. Dependence matrix W for the test case

	P_1	P_2
P_1	0	0,25
P_2	4	0

On the basis of data available in real-time and matrices prepared in the modeling and implementation stage, the system calculates the value of the aggregate indicator:

$$KPI_{Z_1(P_1)} = 0,25 * 100 = 25 \text{(suggested number of ambulances for } P_1) \quad (9)$$

$$KPI_{Z_2}(P_2) = (1 - 0,25) * 100 = 75 \text{(suggested number of ambulances for } P_2) \quad (10)$$

The above calculations for KPI_Z support the decision-making process for the allocation of resources. They represent suggestions to the operator regarding the allocation of resources.

5   Summary

The paper focuses on the current issues of the use of KPIs as mechanisms to support the management of an intelligent urban agglomeration (Smart City). The work was based on conclusions pointing at the shortcomings of simple KPIs, built-in as standard in many information systems that support decisions. Such indicators are independent of each other. Thus they constitute information silos.

The quality of decisions taken in any organization - also in a city - depends on the data available to decision-makers and the context of their use. In the studied case such a context are the processes of the city and every decision has to be taken with regard to their conditions and limitations. In the silo KPI approach, however, such context does not exist. The decision support model proposed in this paper is the answer to this problem. It introduces KPI_Z mechanisms, or aggregate KPIs. Such indicators should be defined in the context of all procedures, by which data critical to the decision maker is aggregated. The data that the model takes into account during the calculation of such an integrated context are associated with each procedure by initiating (KPI_I) and supporting (KPI_W) indicators and the matrices described in the previous section.

The presented model is a simplification. It assumes only those situations where two phenomena occur simultaneously generating potential conflict regarding resources. Hence a two-dimensional priorities matrix is proposed. Anticipating the need to simultaneously launch n procedures, this matrix would need to be given n dimensions. This complication, however, forces the expert team to identify and develop a much larger number of dependencies between processes. But this raises similar concerns to be faced by a knowledge engineer in the search for knowledge - should the knowledge base be complete (such a goal is time-consuming and difficult)? Perhaps, conversely, it should contain only knowledge sufficient to make most decisions (but easier to save)?

These questions result in another goal of research for the authors. That is an extension in the model of the possibilities offered by the dependence and priorities matrices. In the presented concepts they contain numerical values that indicate the strength of relationships between KPI_I and KPI_W indicators respectively and the procedures P themselves. These relationships in complex organizations, however, might not be described so clearly. Therefore, the model should give the possibility to include certain conditions in the matrices beyond the numerical values themselves. These conditions are simple IF-THEN decision rules (implications), which the authors are planning to include in the model as elements influencing the results generated by the aggregate KPI_Z.

Aggregate KPI_Z indicators should be applied in any organization where the number of operating processes is large and at the same time they are linked together with the resources necessary to implement them. The above plans to develop the model are a response to such a need. At the same time, they illustrate the complexity of the problems of managing a Smart City as seen through the prism of the city processes.

References

1. IBM Intelligent Operations Center 1.6: Defining Data Sources. IBM (2013)
2. IBM Intelligent Operations Center: KPI Implementers Guide version 1.0 (2011)
3. IBM Intelligent Operations Center Website (2014). http://www-03.ibm.com/software/products/en/intelligent-operations-center/
4. Central Statistics Office: Prognoza ludności na lata 2014-2050 – report, Warszawa (2014)
5. Deakin, M. (ed.): Smart Cities :Governing, Modelling and Analysing the Transition. Routledge, New York (2013)
6. Campbell, T.: Beyond Smart Cities: How Cities Network, Learn and Innovate. EarthScan, New York (2012)
7. Parmenter, D.: Key Performance Indicators: Developing, Implementing and Using Winning KPIs. Wiley, Hoboken (2007)
8. Parmenter, D.: Key Performance Indicators for Government and Non Profit Agencies. Wiley, Hoboken (2012)
9. Hatzelhoffer, L.: Smart City in Practice: Converting Innovative Ideas Into Reality: Evaluation of the T-City Friedrichshafen. Jovis, Berlin (2012)
10. Klaster Smart IT (2014). http://smartpl.org

Smart Cities System Design Method Based on Case Based Reasoning

Cezary Orłowski[1(⊠)], Artur Ziółkowski[1], Aleksander Orłowski[2],
Paweł Kapłański[2], Tomasz Sitek[3], and Witold Pokrzywnicki[3]

[1] WSB University in Gdańsk, Gdańsk, Poland
{corlowski,aziolkowski}@wsb.gda.pl
[2] Gdansk University of Technology, Gdańsk, Poland
{aleksander.orlowski,pawel.kaplanski}@zie.pg.gda.pl
[3] Staples Advantage Poland SP. Z O.O., Gdańsk, Poland
tomasz.sitek@staples.com,
witold@urtico.com

Abstract. The objective of this paper is to present the results of research carried out to develop a design method for Smart Cities systems. The method is based on the analysis of design cases of Smart Cities systems in cities, the selection of the city appropriate to the requirements for implementation and application. The Case Based Reasoning method was used to develop the proposed design methodology, along with mechanisms of the conversion of project processes and roles to Rational Unified Processes (RUP). The prerequisite for the proposed method is that the enterprise manager must be knowledgeable about high-level Smart Cities system architecture and the design framework applied. The authors, being themselves knowledgeable about architecture of this kind and about project environments which implement KPI models, propose a generic solution applicable to any environments and system architectures.

Keywords: Smart cities · Case based reasoning method · Decision support systems · Knowledge management

1 Introduction

Due to the complexity of their functions and because of the existing social pressure to ensure comfortable living standards for their citizens, today's agglomerations are aiming towards creating management structures known as Smart Cities. These structures are generally defined as the systems which use Information and Communication Technologies (ICT) for the purpose of augmenting the effectiveness of the city's own resources and its component factors in order to improve the living standards of residents [4].

The process of applying ICT for the design of these structures is a complex one, due to the number of technologies used simultaneously. It is also a long-term process, because of the need of the iterative application of these technologies, and it involves the use of significant – and usually distributed – human, hardware and software resources [1, 2]. Thus, the design of ICT systems of this kind usually represents projects charged

© Springer-Verlag GmbH Germany 2016
N.T. Nguyen et al. (Eds.): TCCI XXV, LNCS 9990, pp. 43–58, 2016.
DOI: 10.1007/978-3-662-53580-6_4

with considerable execution risk, which require project management teams to make difficult decisions concerning the selection and application of the already mentioned resources. The analysis of IT-related projects where the purpose was to design systems for Smart Cities, indicates three aspects critical for the success of the project: the design method, the management of the teams that are implementing the selected design method, as well as the methodology used to manage the ICT technologies used for the design processes applied under the selected method [1, 4].

The application of a specific design method depends primarily on the first of the management processes mentioned – managing the teams that participate in implementing the selected design method. However, there is often a problem of the lack of stability of project teams in the context of the long-term duration of the project and, additionally, varied levels of knowledge among office personnel involved in the project, not fully convinced that the project will be a success. An adjacent problem is also the need to maintain continuous (as opposed to temporary) collaboration with specialists from multiple domains (office personnel and system designers).

Much can be said, relatively, concerning project teams with regard to their maturity in terms of CMMI (*Capability Maturity Model Integration*), ITIL (*Information Technology Infrastructure Library*) [10, 11], or COBIT. When designing Smart Cities systems, a question emerges as to how large the team is to be, who in the team will represent the interests of the clients, and to what extent a measurement of team maturity makes sense in the long-term perspective. One may also accept subjective measurements of evaluating maturity, which aggregate the experience, knowledge and implementation method of production and management processes, such as the model of the 'maturity capsule' proposed in a previous paper [8].

In the second case of the selection of information technologies it is necessary that at the process design phase one takes into consideration the fact that information technologies change over time along with the way in which they are applied to ensure the integration of the design frameworks and tools used for system design processes [13, 14]. Quite a lot has been written about technology management processes, but the subject of the integration of design technologies and IT resources in Smart Cities systems is not widely known. Should these technologies be limited only to monitoring events or should they offer a partial functionality, to not only provide definitions of events and incidents (groups of events) but also determine key performance indicators? One can also assume a full scope of system functionality, which means that, apart from the previously mentioned functionalities, the system should also generate notices, alerts and event rules, thanks to the creation of Standard Operating Procedures (SOP), and it should be capable of analyses at the Business Intelligence level [5, 6].

The answer to these questions lies in the correct selection and application of integration technologies when designing Smart Cities systems. If we take into account Service Oriented Architecture (SOA) design with an integration bus, data warehouses or ordinary integration of databases, then (with product information) we possess a compendium of knowledge about integration technologies [2, 4]. Unfortunately, they require comprehensive changes in the cities systems and the introduction of technology with full awareness of the city's needs with regard to these systems. The previous experience of the authors indicates that this knowledge is limited and, given the constantly changing conditions in the functioning of urban environments, it fails to create the conditions for a full-scope design of systems like these.

In today's approach to Smart Cities systems design, the prevailing technology is the 'top-down' method, borrowed from the design mode used for corporate architecture, where the city is analyzed as an entire organization, and then SOA architecture is designed, and specific services are defined. The 'bottom-up' method is used relatively rarely, which is a result of designing individual systems for cities without the vision of the High-Level Architecture of the Smart Cities systems. In many publications mixed-approach solutions are presented, where city processes are treated from the perspective of enterprise architecture and the events and incidents are perceived from the organizational silo perspective of city departments. To sum things up, there is a lack of a uniform (reference) method for designing Smart Cities systems, whereas the majority of cities strongly emphasize the need to design these systems in a compre-hensive way, due to the long-term design perspective. In view of all this, the authors of this paper propose a generic method which may be used as an organization process component for the requirements of any city.

Therefore, in these days when architectures are based on microservices and the microservice design process is supported by domain ontologies [3, 7, 8, 9] complete knowledge of the city system architecture is not required, and the design process can be carried out in an iterative way. This approach would not be possible with the high-level approach to system architecture, without the need to take into account the SOA approach.

2 Analysis of the Case Based Reasoning Approach to Process Design

The characteristics of the methods used for designing Smart Cities systems, described above, indicate that the majority of large cities depart from the approach based on Enterprise Architecture and SOA [13, 14]. The reasons for this lie in the difficulties with a consistent and comprehensive definition of city requirements with regard to these systems. At the same time, talks with project managers responsible for the implementation of Smart Cities systems indicate a considerable pool of experiences which could be used by other project managers in other cities pursuing similar projects. This assessment indicates a need for the analysis of specific design cases, in order to try to generalize them so as to generate a consistent and/or case-based specific method of designing Smart Cities systems for the requirements of a specific city.

With the aforementioned considerations in mind, the authors of this paper have proposed a solution based on case analysis, with the proposed processing approach being Case Based Reasoning (CBR). CBR is a method widely used when trying to generalize cases or when selecting cases to support processes. A relatively large number of cases are about the application of the CBR method to support design works [12]. The analysis of these works prompted the authors of this paper to suggest the idea of using CBR in the process of building the design method for Smart Cities systems.

This method is based on a sequential process of an analysis and selection of cases for the needs of the proposed solution. What is required, is a formal or semi-formal description of the analyzed cases and the use of a consistent notation for both the description of input cases and the description of the proposed solution(s). These descriptions are subsequently stored in the case repository. There are three main formal description methods:

- Representation of cases based on attribute sets
- Object-oriented representations
- Graphical representations

It is also necessary to determine the method to calculate the similarity of cases. In this respect two methods should be mentioned. The first one is the identicalness test (1), and the second, the bottom-limited symmetric similarity test (2) written as the following expressions:

$$S(p_p, p_z) = \left\{ \begin{array}{ll} 1, & dla \quad p_z = p_p \\ 0, & dla \quad p_z \neq p_p \end{array} \right\} \tag{1}$$

$$S(p_p, p_z) = \left\{ \begin{array}{ll} 0 \quad dla \quad p_p \leq p_D \\ 1 - \dfrac{|p_p - p_z|}{\max(p_p, p_z) - p_D} \quad dla \quad p_p > p_D \end{array} \right\} \tag{2}$$

where:
p_p- is the parameter value for the designed case
p_z- is the parameter value for the implemented case
p_D- the bottom limit of the parameter value

The processing algorithm under CBR encompasses four processes (known as 4R). The first process in the analyzed algorithm is *Retrieval* – searching in the Case Base stored in the project repository for the most relevant cases as similar to the analyzed one as possible. To make this possible, it is necessary to define the method by which the similarity of cases is measured. The similarity may be determined on the basis of the tests described above or proposed in the papers [12, 13]. The next step, *Reuse,* involves the application of the closest case to the analysis of the problem at hand. Another step, *Revision,* involves the verification and adaptation of the solution thus obtained to include it in the case knowledge base. The final step of the CBR processing algorithm, *Retainment*, is the process of saving the new/modified case in the CBR Case Base.

3 Requirements Concerning Smart Cities Systems

The first requirements discussed were those regarding Smart Cities systems. They were a consequence of the project carried out for the city of Gdańsku under the Eureka E! 3266 EUROENVIRON WEBAIR project. The purpose of the project is the construction of fuzzy/intelligent decision-making models supporting the management of air quality in agglomerations. At the stage of preparing the application it was not assumed that it would be a Smart Cities system. The system was treated as a transactional system for the processing of pollution emissions data and presenting these data on the city map. It was, however, assumed that the expansion of the existing system will be appropriate in the future, where the specific transport infrastructure of Gdańsk and the impact of its big enterprises (Lotos, Heat and Power Plant, Port of Gdańsk) will be included in the formation of air pollution maps for the agglomeration area. Then the developed solution will become an easily adaptable system for any agglomeration facing environmental protection problems and industrial risks, as well as the consequences of these risks for the residents.

It was also assumed that the WEBAIR system developed under the Eureka project would create the conditions for the creation (by municipal authorities) of long-term investment plans and development strategies. This will be possible thanks to the use of fuzzy scenarios for the agglomeration development and the verification of these scenarios in an integrated IT environment, combining urban and environmental aspects. It was believed that this approach would make it possible to verify the effectiveness of environmental management systems, based on ISO 14000 and EMAS, applied in enterprises, and that it would also allow for the fast identification of possible threats resulting from road transport. Such widely defined requirements provided the basis of the search for a project framework which would create the conditions for designing a system meeting these requirements.

This is why, prior to commencing the development of the system for the City Council, a decision was made concerning the selection of the architecture for the future system. Taking into consideration the advantages of an Enterprise Service Bus (ESB), it was decided that the solution would be SOA supported by ESB. Next, the requirements regarding the future system were determined, data architecture was designed based on the MS SQL environment, and application architecture was designed based on the Rational Software Architect (RSA) environment. The integration bus was built on the basis of the WebSphere Message Broker Toolkit. The presentation layer was presented in an Intelligent Operations Center (IOC) system. The primary advantages of this solution are: dynamic data transformation and conversion, distributed communication, and intelligent service routing (Fig. 1).

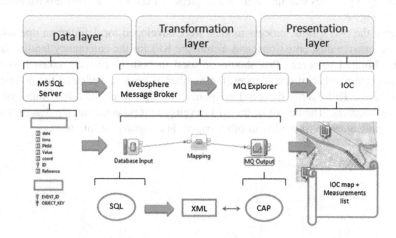

Fig. 1. System integration bus ensuring data flow to an IOC system

It has relatively quickly become known that even though it is possible to build a system based on SOA, a comprehensive view of city processes and their implementation in IOC is rather unlikely. The reasons for this situation are: the risk related to such a comprehensive implementation (and the city is not ready to take on this risk) and a lack of conviction among the decision-makers that it is necessary to implement such a system. Therefore, the proposed design method for the Gdańsk Smart Cities system (for a comprehensive implementation of IOC) was the bottom-up method instead for the top-down method. The bottom-up method is based on an iterative and incremental approach. In order to create the conditions for the reuse of this method (i.e. use by other cities interested in IOC design), the CBR approach was proposed for defining the guidelines for the design method and its reuse for other cities.

The proposed approach is based on the experiences of the CAS Gdańsk design team and it includes an iterative approach to both defining the guidelines of the city processes and to the selection of design tools and design components for the needs of the design process.

The starting point for defining the method is to gather experiences during designing the system for Gdańsk (dust and noise records, and dust and noise simulation models, as well as KPIs for dust and noise emissions) in the first project phase. Because the first project phase ends with the delivery of the operating version to the city during a project meeting, the city decision-makers expect a response to the question of whether a positive case of designing only silo-type solutions concerning dust and noise may be translated into the implementation of other silo solutions for the city. Taking into consideration this question as well as the needs of other cities (Szczecin and Koszalin) concerning IOC design, experimental design works were carried out on an analysis of the possibility of designing any type of silo IOC functionalities. The work was done by a team of six people, managed using the SCRUM methodology plus best practices. The choice of SCRUM, including the duration of the two sprints, resulted from the high level of maturity of the project team and collaboration with another team working on the implementation of KPI models for the needs of the Business Monitor. IOC indicators were defined within the framework of 31 design processes. The implementation of the design processes was modeled as a sequence of tasks in the Business Modeler, as shown in Fig. 2.

During the two sprints, process models were developed for defining an operational KPI for excessive dust emissions, and a strategic KPI for the emergency landing of an airplane. Limited resources that should be used in both cases were taken into consideration (the number of ambulances, the number of hospital beds and the number of fire brigades). KPIs were designed to notify the operator about the status of the available resources. The details of the KPI development process, and the description of the two guidelines are provided in other papers accompanying this work.

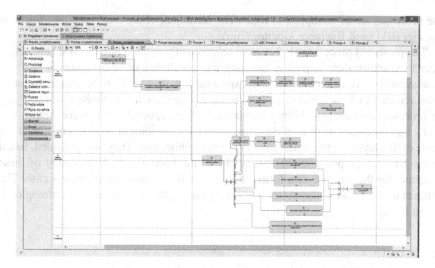

Fig. 2. Method of the implementation of design processes

4 Case Analysis of Designing Smart Cities Systems for Gdańsk

Based on the requirements presented above, the project team developed a sequence of design processes for the delivery of the requirements presented in the previous section. The team did not manage to enter the design processes into the SCRUM product backlog nor did they create a project schedule. The method of designing and setting the priorities for individual design tasks was a result of the experience and expertise of team managers rather than of the best practices of SCRUM or PRINCE 2 and/or RUP (Rational Unified Process). Therefore, the description of design processes presented below represents an ad post set and provides the basis for process generalization, which in turn was the basis for formulating the Smart Cities design processes. In the processes of generalization, attention was paid to the processes unified in RUP as well as to the defined roles.

When analyzing the design processes for IOC, attention was paid to the inter-penetration of products delivered in the design process and the supporting processes and roles. Therefore, it was proposed that the design processes be split into two phases. The first phase included the mapping of IOC processes and roles to RUP processes. In the second phase, these processes were aggregated, according to the layers of the high-level architecture of the Smart Cities system. This architecture was developed on the basis of experiences in designing Smart Cities systems, the requirements of the city, knowledge of technology, and expertise in designing complex systems. This was necessary for the project manager and it seems that knowledge about this architecture is essential, regardless of the team management methodology applied.

Below, the design processes are shown, based on the timeline along which they appear. Next to the design process, the process identifier is provided as well as the identifiers of the project team members who are responsible for process implementation.

The processes are also mapped to RUP. The identifiers of the tools used are: Business Modeler (BM) and Business Monitor (Bm).

The presented processes (1–31) were subsequently entered into the BM (Fig. 2), in order to visualize their sequence, meaning and possibility of being assigned business measurements. At this phase of the experiment it was difficult to assign business measurements to these processes, therefore the presentation was limited to displaying the processes and the possibility of their shared verification by the project team.

Design processes were then classified from the viewpoint of product creation at the levels of five layers of the Smart Cities systems architecture. They also provided, based on the KPI design processes shown in Fig. 2, the basis for the aggregation of these indicators for the purpose of a formal description of the production process case. The architecture of Smart Cities systems for designing KPIs with the use of ontologies is shown in Fig. 3.

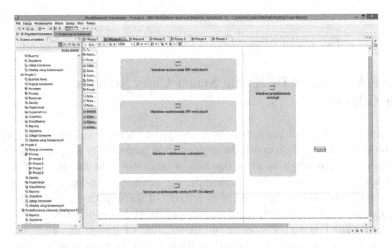

Fig. 3. Aggregated model of design processes

Five process layers were identified and they were assigned to the production of products for five layers of the Smart Cities systems architecture: building simple KPIs for data (processes 1–3), guideline modeling (processes 4–6), building KPI models for the guidelines (processes 7–18), designing ontologies for city processes (processes 19–22) and designing guidelines for KPIs (processes 23–31). The choice of these five process layers resulted from the experiences of the project team, managed with the use of the SCRUM methodology and best practices drawn from other project management methodologies. The application of SCRUM is possible only when the Product Owner knows the High Level Architecture and defines the design processes on this basis.

The selection of these five layers was a consequence of the construction of IBM WebSphere design tools and their philosophy of integration for the needs of the KPI design process. It should be mentioned here that the creation of KPIs represents an integral part of designing systems for companies. It seems that IBM used the experiences of design tools for modeling KPIs for companies, for the needs of cities.

Managing a city with the use of KPIs is still a subject of debate, yet it is the subject of one of the papers accompanying this work, where city processes and formal objects are analyzed in order to demonstrate to what extent a KPI-based solution may be used for the needs of the management of smart cities.

5 Model Case of Designing Smart Cities Systems for Gdańsk

The gathered design processes, as well as the experiences related to the use of CBR, provided the project authors with a basis to define a case for Gdańsk, which will provide the basis to enter this case into the Case Base and later analyze it. In the preparation of the Case Base a generic description was used in which processes were mapped using a matrix description. The description and the treatment of aggregated processes as a system with a feedback loop is shown in Fig. 4.

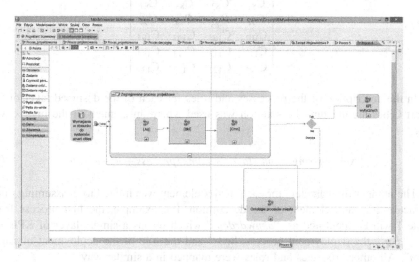

Fig. 4. Generic description of Smart Cities systems design processes, based on CBR

The aggregated design processes and roles have been saved as a matrix of processes and roles:

$$\mathbf{A} = \begin{bmatrix} A_{ij} \end{bmatrix} \quad \mathbf{B} = \begin{bmatrix} B_{kl} \end{bmatrix} \quad \mathbf{C} = \begin{bmatrix} C_{mn} \end{bmatrix} \tag{4}$$

where for the analyzed case of the city of Gdańsk

$$i = 1, 2, 3; \; j = 1, 2; \; k = 1, 2, 3; \; l = 1, 2; \; m = 1, 2, 3, 4, 5, 6, 7, 8, 9, 10, 11, 12; \; n = 1, 2, 3, 4$$

$$A = \begin{bmatrix} A_{11} & A_{12} & A_{13} \\ A_{21} & A_{22} & A_{23} \end{bmatrix} \tag{5}$$

$$B = \begin{bmatrix} B_{11} & B_{12} \\ B_{21} & B_{22} \\ B_{31} & B_{32} \end{bmatrix} \tag{6}$$

$$C = \begin{bmatrix} C_{11} & C_{12} & C_{13} & C_{14} \\ C_{21} & C_{22} & C_{23} & C_{24} \\ C_{31} & C_{32} & C_{33} & C_{34} \\ C_{41} & C_{42} & C_{43} & C_{44} \\ C_{51} & C_{52} & C_{53} & C_{54} \\ C_{61} & C_{62} & C_{63} & C_{64} \\ C_{71} & C_{72} & C_{73} & C_{74} \\ C_{81} & C_{82} & C_{83} & C_{84} \\ C_{91} & C_{92} & C_{93} & C_{94} \\ C_{101} & C_{102} & C_{103} & C_{104} \\ C_{111} & C_{112} & C_{113} & C_{114} \\ C_{121} & C_{122} & C_{123} & C_{124} \end{bmatrix}. \tag{7}$$

In the description of the processes and roles for each of the designed layer of the Smart Cities architecture the solution was used where an element of the matrix was e.g.:

$$[A_{ij}] = string(name\_layer + name\_process + name\_role) \tag{8}$$

The number of characters for each matrix element was limited to 9, assuming three characters for each element of the description. For example, the first process from Table 1 will be identified as: *"dkpwdrdem"* which means a simple layer of KPIs for data *"dkp"*, the deployment process *"wdr"* and the role of the deployment manager *"dem"*. All other processes and roles were mapped in a similar way.

Afterwards, these matrices were mapped to RUP roles and processes (Fig. 5)

$$A_{RUP} = [A_{ij}] \quad B_{RUP} = [B_{kl}] \quad C_{RUP} = [C_{mn}] \tag{9}$$

where for the analyzed case of Gdańsk the number of processes and roles for each group was defined on the basis of Table 1.

$$i = 1, 2; \ j = 1, 2; \ k = 1, 2, 3; \ l = 1, 2 \ m = 1, 2, 3, 4; \ n = 1, 2, 3, 4$$

Thus, RUP processes and roles provided the basis to define processes and roles for the Smart Cities systems design methodology.

$$A_{RUP} = \begin{bmatrix} A_{11} & A_{12} \\ A_{21} & A_{22} \end{bmatrix}$$

Table 1. Mapping of IOC processes and roles to RUP processes

Item no.	IOC process	IOC role	RUP process	RUP role
1.	Deployment of a presentation system for dust and noise data	Support team	Deployment	Deployment Manager
2.	Developing basic KPIs for dust and noise (data)	Support team	Implementation, testing, deployment	Tester Designer Implementer
3.	Developing SOP procedures for dust and noise (data)	Support team	Implementation, testing, deployment	Tester Designer Implementer
4.	Developing by experts two guidelines for managing smart cities: procedure guidelines for operational data – dust, for monitored data; and strategic process for an aircraft emergency landing	Domain experts	Business modeling	Business Process Analyst
5.	Installing the integrator	Technology expert	Implementation	Implementer
6.	Building the models of the two guidelines – BM	Technology expert	Analysis and design	Designer
7.	Modification of both KPI models in BM	Technology expert	Analysis and design	Designer
8.	Building the decision KPI	Technology expert	Analysis and design	Designer
9.	Automated e-mail generation from the KPI level	Support team	Implementation, deployment	Implementer Deployment Manager
10.	Automated SOP generation from the KPI level	Technology expert	Implementation, deployment	Implementer Deployment Manager
11.	Switching the BM version to advanced	Technology expert	Implementation, deployment	Implementer Deployment Manager
12.	Converting the KPI model from BM to Bm	Technology expert	Implementation, deployment	Implementer Deployment Manager
13.	Modification of KPI models from BM	Technology expert	Analysis and design, Implementation, deployment	Designer Implementer Deployment Manager
14.	Modification of KPA [KPI] models in Bm	Domain experts	Business modeling	Designer
15.	Developing a basic KPI to verify the migration from BM to Bm	Technology expert	Analysis and design, Implementation, deployment	Designer Implementer Deployment Manager
16.	Entering data into Bm	Support team	Deployment	Deployment Manager
17.	Supplementing the guidelines with support team data for the needs of Bm	Domain experts	Business modeling	Designer

(Continued)

Table 1. (*Continued*)

Item no.	IOC process	IOC role	RUP process	RUP role
18.	Creating ontologies to aggregate entities from both guidelines by an ontology expert	Ontology expert	Business modeling	Designer
19.	Ontology modification	Domain experts	Business modeling	Designer
20.	Modification of guidelines following the introduction of the ontologies	Domain experts	Business modeling	Designer
21.	Installing the new version of BM	Technology expert	Implementation, deployment	Implementer Deployment Manager
22.	Displaying the ontology in an *. xml file with the concept vocabulary for the verification of guidelines	Ontology expert	Business modeling	Designer
23.	Building collective KPIs	Technology expert	Analysis and design, Implementation, deployment	Designer Implementer Deployment Manager
24.	Naming the basic KPI	Domain experts	Business modeling	Designer
25.	Modification of the newly-named KPI	Technology expert	Analysis and design, Implementation, deployment	Designer Implementer Deployment Manager
26.	Aggregation of processes for the needs of KPIs	Technology expert	Analysis and design, Implementation, deployment	Designer Implementer Deployment Manager
27.	Assigning instances for the needs of the aggregated processes (ET)	Technology expert	Analysis and design, Implementation, deployment	Designer Implementer Deployment Manager
28.	Determining KPIs for three aggregated processes: ambulances, dust and airplane	Technology expert	Analysis and design, Implementation, deployment	Designer Implementer Deployment Manager
29.	Sending three KPIs to Bm	Technology expert	Implementation, deployment	Implementer Deployment manager
30.	Implementing KPIs in Bm	Technology expert	Implementation	Implementer
31.	Verifying both models designed in BM and Bm thanks to using the integrator (ED)	Domain experts	Deployment	Deployment Manager

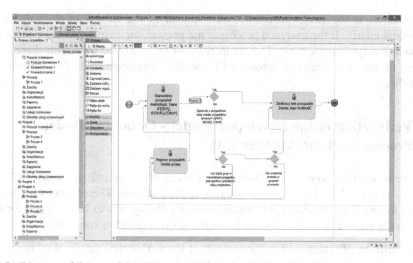

Fig. 5. Diagram of the use of the CBR methodology for the description of a design case mapped to RUP

$$\mathbf{B_{RUP}} = \begin{bmatrix} B_{11} & B_{12} \\ B_{21} & B_{22} \end{bmatrix}$$

$$\mathbf{C_{RUP}} = \begin{bmatrix} C_{11} & C_{12} & C_{13} & C_{14} \\ C_{21} & C_{22} & C_{23} & C_{24} \\ C_{31} & C_{32} & C_{33} & C_{34} \\ C_{41} & C_{42} & C_{43} & C_{44} \end{bmatrix}$$

The matrix description takes into consideration the number of processes and the number of roles responsible for the execution of these processes. For example, the processes of building simple KPIs for data (processes 1–3) take into account three processes and two roles. The matrix description is necessary for the identification of processes and roles. If we assume that the CBR process is launched, the sequence of the case selection process will follow the proposed matrix description, as presented in Fig. 5.

We consider a case of design process methodology described by aggregated processes written as a case.

$$\mathbf{A_{RUP}'} = [A_{i'j'}] \quad \mathbf{B_{RUP}'} = [B_{k'l'}] \quad \mathbf{C_{RUP}'} = [C_{m'n'}] \tag{8}$$

We also assume (Fig. 5) that for any of the matrices the number of elements does not have to be larger than zero (the number of elements is derived during the design process, and this means that a specific design process does not have to occur in each case) and then we start the case knowledge base search procedure. Whenever we encounter a case where all three matrices are present, we investigate the elements of these matrices. If any of the matrices is not present, the analyzed case requires modification. In the search procedure, we also analyze the number of elements of an individual matrix. If the analyzed process is not present, the case needs to be modified by adding a specific element. The comparison of the analyzed case with those contained in the knowledge database will use the identicalness test described by algorithm 1.

6 Verification of the Developed Model - Obtaining Design Process Data for Another City

The basis for the model verification were documents delivered by another city in Poland, interested in the implementation of an IOC system. These documents represented the basis to determine the requirements for the system to be designed for the city. In the requirement analysis the following documents were considered:

- Description of threats and an assessment of the risk of their occurrence, including those concerning critical infrastructure. In this document the category of threat was determined, the likelihood of its occurrence, its consequences and methods of action should the threat materialize.
- Safety net – a document which presents threat categories and persons who should be notified the moment these threats occur.
- Action launch – a document describing procedures of conduct the moment a threat occurs.
- Short-term action plan – a document in which procedures of conduct are presented, but only for specific threats such as dust and noise.

The set of documents presents, on the one hand, a formal description of threats and actions launched when these threats occur and, on the other hand, a formal description of the requirements set for the system. Based on the requirements, design processes were worked out, necessary for developing a system for the needs of the city.

Because the City Council which presented these requirements expected a proposal of the methodology for designing an IOC system, it was suggested that the case of designing IOC for Gdańsk, saved in the Business Modeler, be used for the city which provided these requirements. The simplest requirements (concerning dust and noise emissions) were included in the "Short-term action plan" and, therefore, these requirements were analyzed with a view to using a similar method of system design. Similarities and differences between the system requirements in Gdańsk and those proposed by the city were assessed. It turned out that, despite the similarity of procedures (regulations concerning PM10 are similar for all cities in Poland), the method of implementation of these procedures on the basis of the "Safety net" is different. The designing of KPIs is also different.

At the same time it was determined that the set of four documents provided deliver a full picture of threats, procedures of conduct and decision-makers, which makes it relatively easy to define ontologies for individual KPIs, KPI groups and the entire system. Therefore it seems appropriate to include in the ontology layer the processes related to the preparation of the full picture of threats and their significance, as is the case in the city which provided the requirements. Thus, despite the existing process of creating ontologies for individual KPIs, the process of creating ontologies for all city processes will be added to the ontology layer.

The authors had certain doubts concerning the placement of aggregated ontology processes in the feedback loop, as was done in the process model in Fig. 3. From the viewpoint of both the model (Fig. 2) and the presentation of objects that create the case for CBS, the role of the feedback process is not completely clear. Attempts to determine this were made through verification procedures. Because verification procedures provided information on the significance and role of this process, both the object (ontologies of city processes) and its links with other processes were kept, assuming the key importance of the ontology layer for the design process. It was also determined that a modification of two processes would be appropriate for the simple KPI design layer and the KPI model. The changes will be entered into the case base for designing Smart Cities systems.

7 Conclusions

The paper presents the method of designing Smart Cities systems based on case analysis. The presented method becomes effective in the situation of problems related to designing systems of such complexity, where there is no possibility to use both soft and hard project management methodologies. It is based on the analysis of cases, the conversion of the saved design processes and their use again for designing systems in any city. Thus the case base becomes a knowledge base in which methods of designing systems for any cities are documented.

The paper has demonstrated that building such a knowledge base allows the identification of design processes and design roles for designing specific architecture layers for Smart Cities systems. The one presented in the paper is a consequence of IBM technologies used in design processes. Those used in this work for the modeling of processes and their integration, present the status of the city in the form of KPIs. It should be noted that these indicators, typical for organizational management, were adapted for the needs of smart cities management, and thanks to this it was possible to use these indicators in designing architecture for Smart Cities.

Indicator design processes were carried out on the basis of organizational design processes. To bring them closer to being realized a research experiment was carried out, where a project team was created with the participation of actual system design specialists. The experiment, the purpose of which was to work out a methodology through the selection of two city management cases and later verification based on the requirements of another city, demonstrated the applicability of the proposed method.

An essential role in designing the systems and supporting the high-level architecture was played by the ontologies of city processes. In the case of Gdańsk data, the process of the creation of ontologies was a complex one (limited quantity of data). On the other hand, in the case of another city interested in the design process, a large volume of documents brought order to the process of designing the ontologies, and equally brought order to the design guidelines and the use of cases saved in the knowledge case base.

Using Case Based Systems for the processing of cases shows that, in a situation where the development of *top-down* systems becomes impossible, and *bottom-up* systems strongly limited, the use of case analysis methodology for designing the processes and selecting the roles and their conversion to the RUP format, and saving them in the knowledge case base creates the conditions for selecting the design methodology for the needs of cities. It should be noted that the use of this methodology is limited by the knowledge of the system's high-level architecture and the design environment. The solution presented in this paper is free of these restrictions because it creates a consistent processing framework and may be applied to any architectures and project frameworks.

References

1. Bhowmick, A.: IBM Intelligent Operations Center for Smarter Cities Administration Guide 5. Event flow diagnostic and validation tool for IBM WebSphere Business Monitor, International Business Machines Corporation (2009)
2. Common Alerting Protocol Version 1.2 (2013). http://docs.osasis-open.org/emergency/cap/v1.2/CAP-v1.2-os.pdf
3. Czarnecki, A., Orłowski, C.: Ontology as a tool for the IT management standards support. In: Jędrzejowicz, P., Nguyen, N.T., Howlet, R.J., Jain, L.C. (eds.) KES-AMSTA 2010, Part II. LNCS, vol. 6071, pp. 330–339. Springer, Heidelberg (2010). doi:10.1007/978-3-642-13541-5_34
4. IBM Intelligent operations center Information Center (2013). http://publib.boulder.ibm.com/infocenter/wasinfo/v6r0/index.jsp
5. IBM WebSphere application server information center (2013). http://publip.boulder.ibm.com/infocenter/cities/v1r5m0/index.jsp
6. IBM WebSphere Broker Message Broker Information Center (2013). http://publib.boulder.ibm.com/infocenter/wmbhelp/v7r0m0/index.jsp
7. Kowalczuk, Z., Orłowski, C.: Design of knowledge-based systems in environmental engineering. Cybern. Syst. Int. J. 35(5–6), 487–498 (2004)
8. Orłowski, C., Kowalczuk, Z.: Modelowanie Procesów Zarządzania Technolo-giami Informatycznymi - Gdańsk: Pomorskie Wydawnictwo Naukowo-Techniczne PWNT, (Automatyka Informatyka). Książka profesorska (2012). ISBN: 978-83-926806-4-2
9. Orłowski, C., Ziółkowski, A., Czarnecki, A.: Validation of an agent and ontology-based information technology assessment system. Cybern. Syst. 41(1), 62–74 (2010)
10. Pastuszak, J., Orłowski, C.: Model of rules for IT organization evolution. In: Nguyen, N.T. (ed.) Transactions on Computational Collective Intelligence IX. LNCS, vol. 7770, pp. 55–78. Springer, Heidelberg (2013). doi:10.1007/978-3-642-36815-8_3
11. Pastuszak, J., Stolarek, M., Orłowski, C.: Concept of generic IT organization evolution model. Zeszyty Naukowe Wydziału Elektroniki, Telekomunikacji i Informatyki Politechniki Gdańskiej 18(8), 235–240 (2010)
12. Pokojski, J.: Zastosowanie metody case-based reasoning w projektowaniu maszyn.WNT (2003)
13. Smith, A.D.: IBM Intelligent Operations Center KPI Implementers Guide for Websphere Software. Document version 1.0 (2010)
14. Snadach, K.: Graphical data presentation in IBM Intelligent Operations Center. Diploma Dissertation. Gdańsk (2013)

Model of an Integration Bus of Data and Ontologies of Smart Cities Processes

Cezary Orłowski[1(✉)], Artur Ziółkowski[1], Aleksander Orłowski[2],
Paweł Kapłański[2], Tomasz Sitek[3], and Witold Pokrzywnicki[3]

[1] WSB University in Gdańsk, Gdańsk, Poland
{corlowski,aziolkowski}@wsb.gda.pl
[2] Gdansk University of Technology, Gdańsk, Poland
{aleksander.orlowski,pawel.kaplanski}@zie.pg.gda.pl
[3] Staples Advantage Poland SP. Z O.O., Gdańsk, Poland
tomasz.sitek@staples.com,
witold@urtico.com

Abstract. This paper presents a model of an integration bus used in the design of Smart Cities system architectures. The model of such a bus becomes necessary when designing high-level architectures, within which the silo processes of the organization should be seen from the perspective of its ontology. For such a bus to be used by any city, a generic solution was proposed which can be implemented as a whole or in part depending on the requirements posed by those cities with respect to the construction of such buses.

The work is divided into four main parts. The first part presents a model of high-level architectural design processes, using ontologies and a data integration bus, which constitutes the generalized experiences of the authors drawn from the design processes of Smart Cities systems. The second part contains a description of the environment in which Smart Cities systems are developed, illustrated with two guidelines and the implementation processes of these guidelines. In the third part, two components of that environment are identified: the data integration bus and the ontologies of city processes. This is done to demonstrate how Smart Cities systems are designed and to show the processes of the permeation of data and the ontologies of city processes in the creation of a high-level architecture. The fourth section contains a description of how the proposed model is applied in the construction of a common integration bus for data and ontologies. The paper summary presents recommendations concerning the applicability of the proposed model.

Keywords: Smart Cities · Ontologies · Ontology driven architecture

1 Introduction

The architectures of IT systems are a major source of knowledge of the processes and data (the structured approach) or the classes and objects (the object-oriented approach) of a designed IT system. Within these architectures, a system is separated into individual layers, of which there are fewer or more depending on the requirements. The Three-Tier Architecture is the most common version of this architecture and it contains

© Springer-Verlag GmbH Germany 2016
N.T. Nguyen et al. (Eds.): TCCI XXV, LNCS 9990, pp. 59–75, 2016.
DOI: 10.1007/978-3-662-53580-6_5

three physically isolated layers: of data, of logic and of presentation. Recently, more and more often Service-Oriented Architectures, as well as those oriented at microservices, are designed, in which each functionality is provided as a reusable service or microservice, available both for the recipient and between the services themselves [3]. The designing of services or microservices depends on the complexity of the system, the strategy of the organization and its tendency to take risks, as well as the availability/lack of availability of an integration bus [4]. It also depends on the degree of complexity of the organization's processes and the possibilities of the implementation of the bus in the framework of IT projects.

Integration buses appeared at the time of the introduction of the Software as a Service paradigm, in which the silo functionalities of an organization - here: the cells and departments, as well as partners - can be integrated [1]. The paper expands the meaning of the integration bus and refers to it not only as a service-oriented platform combining both apps developed on the basis of differentiated technologies, incompatible formats, data resources, and communication protocols, but also as organization processes subject to implementation. The advantages of this solution include, above all, dynamic data transformation and conversion, distributed communications and intelligent routing services. It also seems necessary for the model of the integration bus to be such a generic solution that in the creation of future architectures for future organizations/groups of processes, it is possible to use all or part of the designed services.

In the case of complex systems and their realisation within the framework of IT projects, high-level architectures are designed, which present a picture of the functionalities of the systems necessary for the managers of such projects. The high-level architecture then becomes a set of processes and potential functionalities on the basis of which project tasks are prioritized [7]. The process of assigning priorities relies on the knowledge of such an architecture and allows the project manager to decide which of the functionalities are implemented first, e.g. in the first sprint. Designing such architectures requires either full knowledge on the part of the organization, or the involvement of a business analyst in the design process. The authors suggest that the process of designing high-level architectures should be supported by ontologies and that in the design process, the knowledge regarding the integration bus be used to support the development processes of ontologies and the design of high-level architectures. Ontologies are understood here as a formal representation of the stored knowledge about the organization processes in the form of sets of entities and relationships between them. Thus, creating a self-permeating hybrid allows the high-level architecture design process to have a component-based dimension, independent of the knowledge of business analysts and project managers regarding the state of the organization processes [2].

This paper proposes combining the functionalities of the data integration bus to support high-level architecture design processes with the use of ontologies. Such a combination appeared during the design processes of the Smart Cities system. It was incorporated into the design processes of high-level architectures in projects managed with the use of agile approaches. Due to the architecture created with the help of ontologies, as well as the knowledge regarding the data integration bus, it was possible to generate models of the processes taking place in the organization or the applicable procedures of the city in the design process. Since the integration bus aggregated not only data and the processes of the organization, but also technologies, thus a question arose about how much the knowledge regarding the ingredients of the integration bus affects the design process which uses ontologies. The paper focuses on showing these relationships in order to help the designers of integration buses, as well as those designing ontologies, in the selection of technologies and the use of expert knowledge in the design processes of complex systems.

2 Model of High-Level Architecture Design Processes Using Ontologies and the Data Integration Bus

The starting point for the presentation of the proposed approach is a model of high-level architecture design processes using ontologies and the integration bus. Figure 1 presents the requirements, ontologies of requirements and the data integration bus.

It is through the prism of ontologies that the consistency of the processes can be assessed for their subsequent processing. It is through the prism of the data of the integration bus that the appropriate technologies can be selected for modeling the requirements. Then the requirements and the processes of the organization can be easily represented as a feedback system, in which the ontologies of requirements tune the system from the point of view of the efficiency of the organization processes.

Since high-level architecture provides information only about the organization processes, and designing it requires the use of integrated (using the data integration bus) information technologies, hence the integration bus design process will now be presented. Then it will be possible to generalize the chance of aggregating the design of the integration bus as well as high-level architectures. It is assumed that in designing integration buses, the selection of technology derives from organization maturity, but there is no direct relationship between the maturity of the team and the type of technologies used [8]. Mature teams can in fact use less technologically advanced solutions (software tools or programming languages), but can also use very sophisticated CASE (*Computer Aided Software Engineering*) tools [1]. Similar trends can be observed also among teams with low maturity. Unfortunately, this approach to the use of information technology is reflected in a simple relationship. Those information technologies are applied which the team specializes in and they are changed rather reluctantly. The range of available technologies, however, allows both the choice of types (languages, tools, development environment), as well as specific technologies (Java, .Net, Eclipse, J2EE, etc.).

Information technologies of development used by the development team are reflected in a simple way in the way database technologies are used. While in the case of simple systems the supplier imposes a standard of databases, in the case of heterogeneous

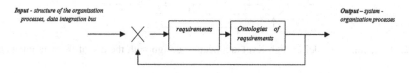

Fig. 1. Data integration bus

systems (varied standards of databases) imposing a uniform standard can significantly hinder both the design process as well as the development of the system. Therefore, the use of data warehousing as a method of data integration is increasing, and properly designed ETL (*extract, transform, loading*) processes enable the input and processing of arbitrary data into the system. The question is, however, to what extent data warehouses will allow the introduction of any given standards of databases in a changing environment (in the long term) typical in the design of Smart Cities systems. Similar questions arise when determining the functionalities of the designed system. Namely, whether its functionalities are limited only to monitoring events or whether the scope of these functionalities is low (a linguistic measure), covering the defining of events and incidents (groups of events), or it is average, covering the designing of key performance indicators. It can also be assumed that the range of functionalities is high, which means that in addition to those previously indicated, a system should also generate notifications, alerts, and rules of events due to the creation of standard operating procedures (SOP). It should also carry out analyses at the level of *Business Intelligence*.

If we include in the functionality design process, described in this way, the design of high-level architectures, the integration bus, data warehouses, or the simple integration of databases, then (along with information about products) we have a compendium of knowledge about integration technologies as well as the ontologies of the organization requirements. Then what is left is the choice of appropriate technologies for the requirements implementation processes. In such a case, it is possible to generalize the process shown in Fig. 1 to a general design of a high-level architecture with the use of the integration bus. Figure 2 shows such a generalized model. For describing the scope and the values of particular variables, a linguistic description can also be applied preceded by an expert evaluation regarding the specified maturities/heights the technologies belong to.

To describe the model, a scalar description was used (as simpler) and it was treated in the same way as the general model (Fig. 1) was treated before, namely, as a control system with control functions f_t assigned to it.

$$sd_t = f_t(ti_t, tbd_t) \tag{1}$$

where:

f_t – control function (rule-based function proposed in this paper)

sd_t – object output response (the level of usability of a designed system), $sd_t \in \ <1,5>$

ti_t – control regulator maturity (requirements level), $ti_t \in \ <1,5>$

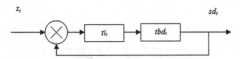

Fig. 2. Generalized model of high-level architecture design with the use of the data integration bus

tbd$_t$ – represents the maturity of the controlled object (the complexity of the designed ontologies), *tbd*$_t \in\ <1,5>$

z_t – setpoint (the structure of the processes and the data integration bus), $z_t \in\ <1,5>$

t - time

Within the work on the generalization of the model, a vector description can also be used taking into account the compendium of knowledge possessed about the built model. Then the complexity vector of the designed ontologies can be expressed as a formula (2)

$$\text{tbd}_t = \begin{bmatrix} mtbd_t \\ ntbd_t \end{bmatrix} \qquad (2)$$

where:

tbd$_t$ - complexity vector of the designed ontologies

mtbd$_t$ - refers to the variable of the ontology design method, *mtbd*$_t \in\ <1,5>$

ntbd$_t$ - refers to the variable of the ontology design tools, *ntbd*$_t \in\ <1,5>$

The full scalar and vector description (partly presented in the paper) creates conditions for the implementation of the model. The function f_t can have its representation in the form of a rule-based function (for both the scalar and vector descriptions). In the case of the scalar description, the rule-based description will include two input variables and one output variable. If the linguistic description applied in this work is assumed, and later the fuzzy one (under consideration for use), the complete model of the high-level architecture design processes will include $2^5 = 32$ rules [12]. An example of the model implementation with the use of a rule-based description is represented by the Eq. (3).

$$\text{IF the requirements level} <ti_t> \text{ is at} <1> \text{ and the ontology complex as}$$
$$<\text{tbd}_t> \text{ is at} <1>, \text{ then the usability of the system} <sd_t> \text{ is at} <1> \qquad (3)$$

The possibility of using both scalar and vector descriptions also means that the general model (Fig. 2) can be used (due to the compendium of knowledge on the status of the integration technology) for the selection of these technologies in the development processes of Smart Cities systems. In this case we have the knowledge about both the types and kinds of given technologies. If, in addition, we are able to present the structure of the model (Figs. 1 and 2), then it can be justified in replicate and also predictively. Therefore, the next stage of the presentation of the model is to demonstrate its validity by presenting the environment in which the validity was researched.

3 Project Environment for Examining the Validity of the Model

The processes of examining the validity of this high-level architecture design model were carried out in a project team working with the IOC (*Intelligent Operations Center*) system. The team is responsible for two research projects (*Eureka and SASD*) whose main objective is to create a system to support decision-making in situations when the levels of particulate matter and noise in Gdansk are exceeded. Hence, the presentation of the project environment was initiated with a discussion regarding the guidelines and requirements of the system in relation to the decision-making systems which are the main objectives of the two projects, as well as regarding the possibility of their implementation. Due to limited access to knowledge on the functioning of all processes, as well as due to limited time, a decision was reached to design a high-level architecture model based on the analysis and implementation of two processes present during emergency situations in Gdansk.

To begin with, the development team used the knowledge of domain experts, and the experience gained in earlier phases of the implementation of both projects to collect requirements in the form of guidelines [10]:

- operational, referring to decisions made on the basis of monitoring the level of air pollution with PM10 (sending ambulances to kindergartens in the event of a significant excess of PM10 levels),
- strategic, concerning decisions made on the basis of a notification about aircraft emergency landing at Lech Walesa airport in Gdansk (sending ambulances to the airport).

The selection of both types of procedures resulted from the need for a decision (the creation of complex KPIs) where resources (the number of ambulances) were limited. While presenting the guidelines, team experts used a document called "Short-term action plan for Gdansk". The document stated two alarm levels. In the case of the first alarm level (24-h concentration of 50 ug/m3), the operator is obliged to inform the following entities:

- crisis management centers of cities,
- the Board of the Pomeranian Region,
- the Provincial Inspectorate of Environmental Protection in Gdansk,
- authorities of the alarmed cities,
- local radio and television.

For the second alarm level (smog alarm) when the concentration exceeds 300 μg/m3, and the value of the 24-h concentration is higher than 200 μg/m3, the operator is required to inform two more entities besides those mentioned above:

- directors of health care facilities and hospitals,
- directors of educational institutions and care facilities.

The representative of the client decided that the implementation will include data regarding the second alarm level, and the crisis management center will take the

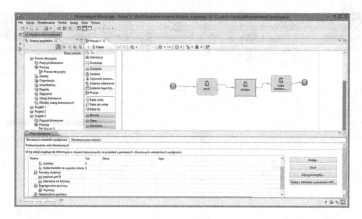

Fig. 3. Example of modeling city processes using the WebSphere Business Modeler

decision to dispatch ambulances to a kindergarten. The second guideline included a procedure which is triggered by the pilot of an airplane approaching the airport when a fault has been detected, such as problems with releasing the landing gear and expected problems when landing. The pilot informs the air traffic control tower, which in turn, depending on the size of the aircraft, calls fire brigades and airport ambulances, and also municipal services. Regardless of the size of the aircraft, the airline and the crisis management center are informed. The representative of the client decided also that standard data regarding emergency landing of an aircraft in Gdansk will be used in the design processes, and the crisis management center will take the decision concerning the dispatch of ambulances to the airport.

Due to the fact that both guidelines involve the same resources (the number of ambulances) in decision-making, for the graphic representation of both guidelines they have been formalized with the use of BPML (*Business Processes Modeling Language*). The design tool called the *IBM WebSphere Business Modeler* Basic version has been used. With the use of tags (similar to the BPML standard) [4, 5], it allows for both the creation of pools which belong to the respective operating units, as well as nodes, loops and conditions (Fig. 3). Such a presentation creates conditions for exporting data to the IOC system. Due to the Business Modeler, it became possible to view the processes creating both guidelines, the responsible entities within these guidelines and the measurement of processes with the use of KPIs.

Knowing the methods of creating the measurements of the processes, the following stage of the design process was to define the required key performance indicators and their presentation in the IOC system. After creating the appropriate measures of KPIs, the limit levels of air pollution with PM10 were defined [10]: 0–49 μg/m3 - appropriate level, 50–299 μg/m3 – warning level for the first alarm level, 300+ μg/m3 - critical level for the smog alarm. In addition, a color was assigned to each indicator: green - no comments, yellow - warning, red - critical. Apart from designing KPIs the development team also launched design processes of standard operating procedures (SOP). Their task was to run the processes of sending e-mail messages to previously defined addressees or starting other KPIs. An example of the KPI defining process is shown in Fig. 4.

Fig. 4. KPIs developed in the IBM Intelligent Operations Center (Color figure online)

After the implementation of four key performance indicators (noise emissions measurement, PM10, the number of available ambulances and an aggregate indicator of the aforementioned three) is was concluded that only the Advanced version of Web-Sphere Modeler rather than the Basic version used so far creates conditions for modeling business processes, which can then facilitate the process of their creation at the level of the Business Monitor. It was also noted that such a use of the Business Modeler each time requires individual mapping of the existing infrastructure, procedures and guidelines, which takes considerable time and resources [6]. Wanting to achieve the greatest simplification of this process, as well as attempting to introduce the principle of reusability of the created components and their modularity, a different approach was considered. It was proposed to save the available knowledge about the city, infrastructure, processes and operational units with the use of ontologies and then use the ontologies in the design processes of the system.

4 Integration Bus and the Ontology of the Environment Processes

The description of the design environment presented in the previous section did not take into account the status of key parameters for verifying the model, such as the status of the data of the integration bus and the advancement status of the description of city processes with the use of ontologies. Therefore, in order to examine the validity of the model, this section begins with a description of the status of the data of the integration bus, to then move on to describing the processes of the city and the environment in which these processes are implemented with the use of an ontology supporting the use of

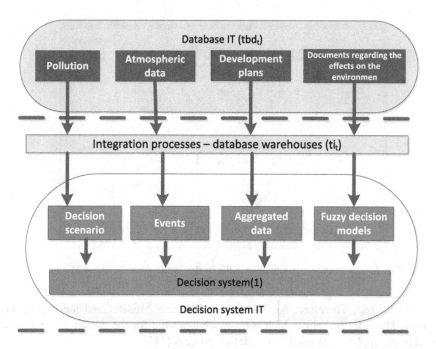

Fig. 5. Data Warehouse as a data integration bus

the integration bus. The starting point for the discussion on the status of the integration bus was the IOC 1.5 *WebSphere Message Broker (WMB)* version of the service bus, where the data undergoing the integration was obtained from many different database resources. A data warehouse developed on the basis of WMB served as the data of the bus. Power processes for both the warehouse and the decision-making system were developed. The standard of the database warehouse was also analysed (with regard to the applicability of the model developed in step 2). Two file standards *.SQL and *.DB2 were considered. In terms of the decision-making system (meeting the requirements of the city) the DB2 database standard proved to be a better solution. However, when the experience of team members who were not familiar with the DB2 standard was taken into account, it was decided that the SQL standard would be used, despite the fact that the DB2 standard seemed more evolutionary [9]. The decision to choose the better-known standard rather than the more prognostic one stemmed from the need to reduce project risk. Figure 5 shows the diagram of data integration processes with the use of the integration bus developed on the basis of WMB (in line with the formal description of the model introduced in Sect. 2 of this paper). WebSphere Business Monitor presented in Fig. 6 took into account the conversion of database resources *. db2, as well as *.SQL to the CAP protocol (common alerting protocol) via *xml*.

The design of the integration bus purely as a data bus stemmed from the experience of the development team. When moving from version 1.5 to 1.6, the WMB tool was replaced by the direct mapping of data without the need to queue at the CAP. Hence, the development team used IOC as an environment ideal for the direct supply of the *.

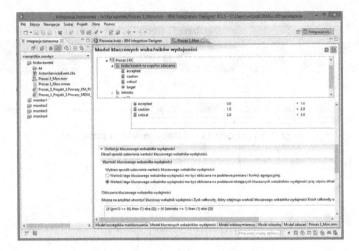

Fig. 6. An example of modeling KPIs in the WebSphere Business Monitor

sql standard data. Therefore, the WebSphere Business Modeler and the WebSphere Business Monitor, the tools which were made available for the design process, were independent applications used in the design process [11].

In such a situation, creating models of the performance indicators with the use of the *WebSphere Business Monitor* was an independent process resulting from installing the *IBM Integration Designer* independent of IOC. This solution enabled the mapping of the existing IT resources in the form of universal service components. In the case of the implementation of indicators conducted by the development team, this tool could only be used as a base for the Business Monitor - the implementation environment of KPIs. Then, the KPI models could be directly exported from the *WebSphere Business Modeler* to the *WebSphere Business Monitor*. As it turned out, this was only possible in the *Advanced* version. It became necessary to incorporate this tool into the development environment. After installing the *Advanced* version and importing the KPI models previously created in the Basic version of the Business Modeler to the *Advanced* version, it was possible to export them to the *.mm format and use them in the Business Monitor for defining KPIs. The example of defining indicators in the *Advanced* version is shown in Fig. 6.

After defining the data of the integration bus, the ontology of the city processes was designed. First, a dictionary of concepts was defined in order to standardize the concepts used and their definitions. Then it was possible to determine the actors, the activities and the assignment of the activities to the actors. The example presented below is of the ontology of guidelines for making an emergency landing (one of the cases analyzed and described in Sect. 3 of designing Smart Cities systems).

Pilot is a notifying − body.
Airport is a notified − body.
Airline is a notified − body.
Airport − Fire − Brigade is − part − of Airport.

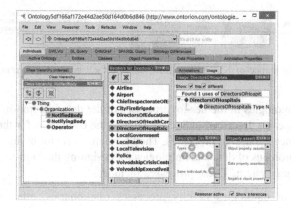

Fig. 7. Window for introducing rules in Protégé

These concepts were saved in the rule base of the Protégé design tool. The example of saving rules is shown in Fig. 7.

Files which were created during saving the above ontologies in the .entcl format, have a purely textual structure, with a basic function for comments. However, they were not consistent with the .mm file structure used by the WebSphere Business Monitor based on XML in which the city processes were saved. To be able to use the ontologies described above to support the description of the city processes, it became necessary to convert knowledge stored in one format to another. To accomplish this, and to support the design of other city processes with ontologies, it was decided to create an own tool called OntoTransTool. A detailed description is given in the paper on ontologies. It then became possible to define a required XML structure, on the basis of a decoded .mm file generated by the WebSphere Business Modeler. With this solution it was possible to translate text rules into the required format. An example of the implementation of the rules using OntoTransTool developed for the design process, is shown in Fig. 8.

```
C:\Windows\System32\cmd.exe                                            _ □ x

d:\OntoTransTool>OntoTransTool.exe --input "SmartCityOntology.encnl" --output "SmartCityModel.xml"
Engine setup...
Loading ontology into reasoner...
Processing Organizations...
Processing Activities...
Setting Up Ownership...
Creating IBM Business Modeller XML File...
Saveing...
...Done

d:\OntoTransTool>_
```

Fig. 8. Example of processing rules with the use of OntoTransTool developed for the design process

Fig. 9. Sequence of KPI design processes and their support with ontologies

The use of ontologies in the design processes of the data integration bus as well as in the design processes described in Sect. 3, created the conditions for the development of generic sequences of design processes. They form the basis for building a model of a common integration bus of data and ontologies for designing the high-level architecture of the Smart Cities system. These processes include:

- the creation of ontologies of city processes (dictionaries, relationships, actors, subjects) and their surroundings.
- the transformation of the saved ontology to define preliminary process models (the main processes illustrating the functioning of an agglomeration)
- the modeling of city processes (detailed processes reflecting the functioning of the departments and the implementation of procedures)
- the creation of KPI models (for data and procedures)
- the presentation of indicators in the IOC system (for data in procedures)

Then the design process of the integration bus (its KPIs) can be presented in the form of sequences of the design processes described above. Figure 9 shows a sequence of KPI design processes in which the city processes are identified, and their analysis, modeling and implementation are supported by ontologies of these processes. Such an approach to processes forms the basis for treating them as components.

The analysis of Fig. 10 shows that three groups of processes and their implementation technologies can be defined in supporting the design process: the modeling of processes for data input, the integration of process modeling technologies and the

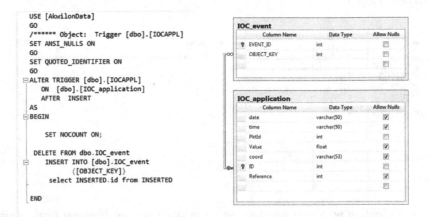

Fig. 10. Akwilon database tables (a) and the data trigger (b), which enables operations on both tables as an example of designing databases with diversified specific fields

input of data and ontologies to support the modeling processes. Such a view on the sequences of processes (in a situation where their technical implementation does not exist in the form of a bus) should be seen by the designer of the system. That was the case during the design of the IOC system (described in Sects. 3 and 4 of this paper) and it formed the basis for the future implementation of these processes as an integrated bus of data and ontologies. It seems that the described sequence of processes can be typical in the design processes of Smart Cities systems and it can form the basis for the design of high-level architectures.

5 Changes in the Ontology of Processes and Further Development of the Data Integration Bus of the Design Environment

The model of the high-level architecture design process with the use of ontologies and the data integration bus (Sect. 2), the design environment to examine the validity of the model (Sect. 3) and the description of the integration bus together with ontologies of the environment processes, which were discussed in previous sections, provided the basis for the development of a model of a common bus of data and process ontologies and its implementation in the design process. Taking into account the implementation processes, the first issue to consider was the possibility of changing the place and role of the ontology and treating it as a service in design processes on the basis of experience gained during the implementation of the solution in Gdańsk. This approach was a consequence of the question regarding the extent to which the adopted model of the designing and implementation of ontologies to support the IOC design processes can be reused. Can the diverse procedures and modeling processes for other cities be supported by previous experiences and services designed on the basis of these experiences? In order to answer these questions, we need to identify the experiences which unite rather than divide in the implementation of Smart Cities systems.

The first group of experiences are those which connect the diverse design processes of a city, which is treated as an organization, with the resources that create it. The diverse structures and resources, such as crisis management centers, departments of government offices, airports, hospitals and educational units can be presented as a system of connections and their hierarchies which is uniform for any given city or varied for particular cities. Hence, it becomes possible to use common or diverse ontologies and treat them as a group of or as individual services supporting the city design process. It is also possible to combine the design processes/ontologies with information technology used for the modeling of processes and implementation of ontologies in order to understand the concept and the importance of these processes. These entities (ontologies, processes, technologies) can also be considered as services used in the design processes and be attached to the integration bus.

The second group of experiences are those which arise from the analysis and the possible use of available city data. Each system designed for the needs of a city is based on monitoring independent data or data which is linked to the bus. The data is used for flows, simple transformations and operations, enabling the standardization and normalization of data, as well as the addition of simple processing procedures. The example application of the integration bus as the data bus is presented in Sect. 4 of this paper. An attempt to aggregate experiences referring to the use of data comes down to the knowledge of the standards used in the implementation of KPIs. The experience gained during designing KPIs suggests the use of *.db2 or *.sql standards or the construction of converters. The construction of mechanisms that convert *.xml to sql proved to be a relatively simple process. Experience acquired in the data usage process focused also on the construction of triggers of processes of acquiring data from external databases to supply the KPIs. An example of the structure of the trigger is shown in Fig. 10a.

The flow of data starts with the *Input* node. It is checked whether there are lines in the *IOC_application* table in the Akwilon2 database containing data from 85 stations measuring noise (Fig. 10). If so, the *IOCAPPL* trigger of new records (*insert* command) is activated to create records with the same ID in the *IOC_Event* table. The *IOC_event* and *IOC_application* tables are linked by the primary key ID. Figure 10b presents a fragment of the code that allows the trigger to conduct operations on rows of both tables.

The third group of experiences focuses on the use of the data of integration buses. Experience from designing the IOC system version 1.5 was used, where it was necessary to create an integration bus based on the WebSpere Business Monitor. Another source of experiences was the generic ESB (*Enterprise Service Bus*) presented by IBM and integrating both infrastructure services as well as development services to support the construction of this bus, access services, business applications or process services. The scope of these services, the connection to the ESB and the sequence of connecting processes is a source of knowledge on how to design such a bus to then design smart cities systems. If it is also assumed that the data of the bus can be the basis for the design of the bus ontology (for streamlining design processes) then a collection of such experiences supported by knowledge on integration buses is the basis for the construction of a bus for designing the systems for smart cities.

Taking into account all the three groups of experiences which are common to most, if not all Smart Cities systems, and which result from the model of the integration bus adopted by the team, it can be assumed that we can create generic models of such buses. Then, the saving of knowledge in the form of an ontology for the purpose of designing a system for another city may be based on the previously created glossary of terms, and it can be extended as a service of an integration bus. The information technologies applied may be similar, which means that the processes and their ontologies can be reusable. Any technologies similar to Protégé, such as FluentEditor and WFLEditor can be used to create a common glossary. It is enough for the mechanism connecting the developed service to the bus to meet the connection criteria with the bus.

Figure 11 shows the construction of the integration bus as a result of experiences which were the effect of design processes and the use of ontologies to support the design of city processes. The figure has been divided into three parts. The upper left section presents data regarding the integration bus which form the basis for the construction of any integration bus (including previously designed integration buses for cities that use them). These include data on the integration mechanisms and the

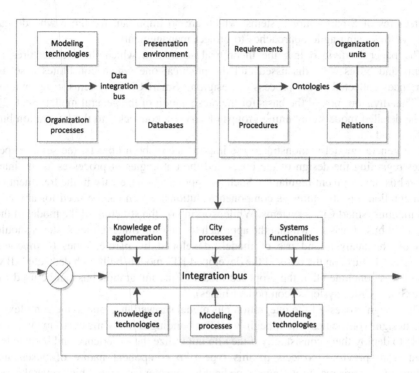

Fig. 11. Model of the integration bus designed on the basis of experiences resulting from the design process and the ontologies of city processes

subsequent use of the modeling technology of city processes, environments for presenting the processes, KPIs, city processes, as well as data and access mechanisms of databases. The upper right section of the figure presents data regarding the structure of the ontology of the city processes to support the design processes. The designing of ontologies is an integrated process of obtaining requirements concerning city systems, data relating to city entities, procedures and relationships occurring both between procedures as well as individuals. The lower section of the figure shows the structure of the bus, taking into account the roles of data and knowledge about the structure of such buses as well as the place and role of ontologies. This hybrid solution results from the work and experience of the team designing Smart Cities systems.

6 Conclusions

The paper is a theoretical and practical study of the structure of integration buses. This combination of theory and practice created conditions for the development of an integration bus model which takes into account both the design experience as well as the ontologies of city processes, modelled with the use of information technologies connected to an integration bus. It was proved that combining the data of the integration bus with ontology processes facilitates designing/seeing the high-level architectures of Smart Cities systems, which are so important for those who design these systems using agile approaches to project management.

The paper is divided into the theoretical section in which the architectures of systems and buses were discussed and the practical one which constitutes a set of experiences collected in the process of designing Smart Cities systems using ontologies. This division formed the basis for the presentation of the general model (Sect. 2), and the detailed models containing groups of services connected to the integration bus (Sects. 3 and 4).

The general model demonstrated the importance of both objects: the set of experiences regarding the design of the buses and the ontologies of processes as the integration bus development regulators. Such an approach is necessary in the treatment of the integration bus structures as component solutions, which can be used for any city implementing Smart Cities systems. While working on the structure of the model of the integration bus, it was shown that the approach to building Smart Cities systems should focus on the analysis and use of models developed by partner cities (component ontologies). In turn, on the basis of the developed KPI models built with dedicated IBM tools, it was concluded that the proposed model of the integration bus can be used for future Smart Cities systems (component buses).

The design process of the integration bus model showed that, due to the complexity of the design work, cities which begin work on building Smart Cities systems will not be able to deploy them consistently if they do not utilize the experience and knowledge gained from previous projects of this type (the component nature of buses and ontologies). For this reason it seems expedient to consider a case in which several cities would implement Smart Cities systems simultaneously (using common data components of buses and ontologies of city processes). Such an approach necessitates cooperation between several cities. Thus, it seems reasonable for the design process of Smart Cities systems to have a bottom-up nature both in terms of designing the integration bus model and in the implementation of the city processes.

The case study of designing the integration bus for Gdansk referred to the modeling processes and the implementation of two guidelines, using the concept of the integration bus in the IOC system. Problems emerging mainly with regard to the modeling processes and the integration of processes and KPIs showed the importance of the bus and the ontologies supporting the design processes. The awareness of this situation demonstrated the need to change the modeling technology and to install an integrator which improved the design processes. The design process became shorter, and the number of errors in defining the city processes was lower. It seems that after the implementation of the integration bus (as in Fig. 11), the design process (through the use of ontologies) will not be burdened with such a significant number of errors and the conversion of the measurement of these processes onto KPIs will not be as redundant as is the case with data conversion from the Business Modeler to the Business Monitor.

References

1. Czarnecki, A., Orłowski, C., Sitek, T., Ziółkowski, A.: Information technology assessment using a functional prototype of the agent based system. Found. Control Manag. Sci. **9**, 7–28 (2009)
2. Czarnecki, A., Orłowski, C.: Ontology as a tool for the IT management standards support. In: Jędrzejowicz, P., Nguyen, N.T., Howlet, R.J., Jain, L.C. (eds.) KES-AMSTA 2010, Part II. LNCS, vol. 6071, pp. 330–339. Springer, Heidelberg (2010)
3. IBM Intelligent Operations Center: Information Center (2013). http://publib.boulder.ibm.com/infocenter/wasinfo/v6r0/index.jsp
4. IBM WebSphere application server information center (2013). http://publip.boulder.ibm.com/infocenter/cities/v1r5m0/index.jsp
5. IBM WebSphere Broker: Message Broker Information Center (2013). http://publib.boulder.ibm.com/infocenter/wmbhelp/v7r0m0/index.jsp
6. Ontology driven architectures and potential uses of the semantic web in systems and software engineering, W3C (2013). http://www.w3.org/2001/sw/BestPractices/SE/ODA/
7. Orłowski, C., Ziółkowski, A., Czarnecki, A.: Validation of an agent and ontology-based information technology assessment system. Cybern. Syst. Int. J. **41**(1), 62–74 (2011)
8. Pastuszak, J., Stolarek, M., Orłowski, C.: Concept of generic IT organization evolution Model. Fac. ETI Ann. Inf. Technol. **18**, 235–240 (2008)
9. Pokrzywnicki, W.: Proces instalacji oraz wdrożenia systemu IBM intelligent operations Center na Wydziale Zarządzania i Ekonomii Politechniki Gdańskiej. Diploma Dissertation. Gdańsk (2013)
10. Stanford Center for Biomedical Informatics Research: Program for the protection of air in the Pomeranian region, Gdańsk (2013). http://www.wfosigw.gda.pl/biura/wfos/article_download/93/10%20-%20Program%20ochrony%20powietrza%20dla%20strefy%20pomorskiej.pdf
11. WebSphere Business Process Management: V7 Production Topologies, International Business Machines Corporation (2010)
12. Zadeh, L.: Fuzzy sets as a basis for theory of possibility. Fuzzy Sets Syst. **1**, 61 (1978). London

Ontology of the Design Pattern Language for Smart Cities Systems

Cezary Orłowski[1]([⊠]), Artur Ziółkowski[1], Aleksander Orłowski[2], Paweł Kapłański[2],
Tomasz Sitek[3], and Witold Pokrzywnicki[3]

[1] WSB University in Gdańsk, Gdańsk, Poland
{corlowski,aziolkowski}@wsb.gda.pl
[2] Gdansk University of Technology, Gdańsk, Poland
{aleksander.orlowski,pawel.kaplanski}@zie.pg.gda.pl
[3] Staples Advantage Poland SP. Z O.O., Gdańsk, Poland
tomasz.sitek@staples.com, witold@urtico.com

Abstract. The paper presents the definition of the design pattern language of Smart Cities in the form of an ontology. Since the implementation of a Smart City system is difficult, expensive and closely linked with the problems concerning a given city, the knowledge acquired during a single implementation is extremely valuable. The language we defined supports the management of such knowledge as it allows for the expression of a solution which, based on best practices recorded in the form of design patterns, is also tailored to the requirements of the city seeking to implement the Smart City solution. The formal/ontological structure of the language in turn allows the automatic management of the properties of a solution recorded in this way. This final feature of the introduced language is extremely important in the decision-making process regarding the choice of a particular solution by the relevant authorities.

The work is divided into five main parts. In the first part we discuss the implementation issue of the integration bus using the example of the IOC. In the next part we talk about the validity of using semantic technologies in order to expand the spectrum of potential implementations. Then we discuss the ontological implementation of the Smart City pattern language which we created, a language which allows for both the saving of requirements and the validation of solutions specified in it. We also present an example of usage, which at the same time serves as a validation of the language in real-life conditions. In the last part we discuss certain aspects of the pattern language and the possible ways to develop research related to it.

Keywords: Smart cities · Ontologies · Semantics · Ontology driven architecture · Design patterns · Controlled natural language

1 Introduction

The concept of a Smart City promises the achievement of a new quality of city management by integrating ICT structures available in the city and by their expansion into other urban areas. The integration of these structures is carried out through the appropriate

© Springer-Verlag GmbH Germany 2016
N.T. Nguyen et al. (Eds.): TCCI XXV, LNCS 9990, pp. 76–100, 2016.
DOI: 10.1007/978-3-662-53580-6_6

implementation of a common, scalable and highly-efficient integration bus, such as the IBM Intelligent Operations Center [2]. Such a bus, being indirectly integrated with the urban space becomes part of it and thus designing the implementation process for the integration bus is highly complicated and expensive. To reduce the costs of further implementations, it should be possible to repeat this process. Due to the diversity of urban areas, such repeatability cannot, however, involve copying successful implementations in full, but rather means the selection of the best combination of best practices, adapted to particular specifics, and then the verification (a priori) of the resulting solution as a whole. The desired selection method of the implementation of the Smart City integration bus should allow for:

1. cataloging of previously developed best practices,
2. verification of the expected results of the implementation.

In this paper, we present a method we developed to manage the Smart City integration bus implementation experience which meets both conditions. It is based on the so-called Smart City pattern language, recorded in the form of an IT ontology. This language is equipped with all language layers:

1. The syntactic layer (and grammar) is provided by a network of possible solutions, which pattern language fits into, patterns become the words in the pattern language.
2. The semantics of the pattern language is defined by the limitations/opportunities of the mutual interlacing (depending on each other) of patterns
3. The pragmatic layer is carried out by appropriately selected tools enabling work with the language. For this reason, we have implemented a key tool (OntoTransTool) allowing for the integration of the semantic technologies with the existing solution for the configuration of the Smart City.

We also describe the verification results of the method using the example of implementing the Smart City integration bus based on the IBM IOC integration bus in Gdansk.

2   The Issue of Implementing the Smart City Integration Bus Based on the Example of the IBM Intelligent Operations Center

In this section, we describe the implementation of the Smart City integration bus based on the example of the IBM Intelligent Operations Center (IBM IOC). The IBM IOC integration bus allows for highly efficient processing of data streams and provides tools for their integration. The implementation of IBM IOC takes place on two levels, a low level - the integration bus level (Fig. 1) and at a high level - the model level (Fig. 2). The two-level process of implementation of IBM IOC, described in this section, is the starting point for extending the implementation process with the ontology level that we propose, which will eventually become the environment to which the proposed Smart City pattern language refers.

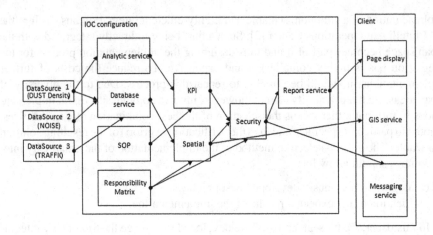

Fig. 1. Sample configuration of the Smart City at the integration bus level (in the M0 MOF layer).

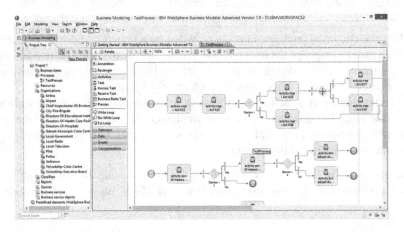

Fig. 2. Sample Smart City configuration at the model level (M1 MOF layer)

The integration bus level allows direct access to the services provided by IBM IOC. Figure 1 illustrates an exemplary implementation of IBM IOC seen from this level. We can see here that the processing of data within the IBM IOC bus is carried out in real time via data streams (DataSource) and the results of processing these streams become the basis for determining key performance indicators KPIs. These indicators are presented to bus operators in the form of an intuitive user interface (Client), both in the form of reports (Report) as well as on a map (Spatial, GIS service) so that they can support them in decision-making. The system also integrates the work of operators through an integrated organization management system, such as workflow (carried out by a set of procedures - Simple Operational Procedure - SOP) and through ICT tools for the direct communication of operators (Messaging Service). A hierarchy of privileges built into the system (Responsibility Matrix) together with the security system (Security) allows the mapping of the actual hierarchy of responsibility.

The integration bus level gives access to the greatest potential of configuration, however, the target configuration is difficult to maintain from the perspective of this level and thus requires skilled IT staff. A configuration of processes such as workflow can be an example here: the flow of documentation and the implementation of the responsibility escalation path at this level requires the entire process to be defined in the form of single SOPs, one of which can launch others by generating chains of launching rules. Amongst all the SOPs, the overall picture of the process is less visible.

In contrast to the bus level, the model level captures the IBM IOC configuration from the perspective of the processes which it is to carry out. It is a natural method for analysts and managers, allowing a certain degree of generality for capturing the holistic aspects of implementation difficult to see from the integration bus level. Modeling at this level takes place in the IBM Business Modeler tool in Business Process Modeling Notation (BPMN) language [14]. The example of the IBM IOC configuration at this level is shown in Fig. 2. This shows both the organizational structure (Organizations), report templates and query definitions as well as (unlike at the bus level) the full structure of workflow processes, together with assigning the relevant participants taking part in them, including messages passing between them.

The model level, however, does not provide full configuration options available from the integration bus level, such as the ability to connect any (external data sources), so the most common configuration of the IOC bus consist of two phases (Fig. 3). The perspective of the integration bus level, generated from the model level, is then supplemented with necessary components and is ultimately implemented at the target IOC infrastructure level. This process takes place according to the rules of model transformation in terms of Model Driven Architecture (MDA). The MDA architecture called Meta-Object Facility (MOF) was introduced in 2001 by the Object Management Group (OMG) and is the result of the evolution of object-oriented modeling methods, in particular based on UML.

Fig. 3. IOC bus configuration process: the transition from the model level through the integration bus level

IBM IOC carries out the MOF architecture. MOF is designed as a four-layer architecture of models, each of which is more general than the previous one. The layers are called: M3, M2, M1 and M0. From the point of view of the IOC, the integration bus level corresponds to layer M0, while the model level to layer M1 (Figs. 1 and 2). The models are essentially networks of interconnected entities, which in turn (the entities) refer to phenomena occurring in a particular domain [4]. Since the models in MOF are represented by graphs, the languages allowing the processing of models are graph transformation languages [16]. In MOF, the transition from the meta-model (a model from a higher layer) to the model occurs via model transformations called Queries/Views/Transformations (QVT) indicated schematically in the figure (Fig. 3).

To sum up: IBM IOC provides the infrastructures and supports the implementation process of the above-mentioned integration bus infrastructure. The support takes place on two levels. At the model level (M1 as referred to in MOF) and the integration bus level (M0 in MOF). During the configuration of the IBM IOC implementation, from the perspective of the model level, we are equipped with a powerful tool for modeling business processes. In addition, at this level we have the ability to model a high-level organizational structure. After the transformation from the model level onto the integration bus level, a full spectrum of opportunities for integration opens up to us.

At this point, it is worth pointing out a serious limitation of the presented method to support the implementation process of the integration bus. There is no possibility to save the knowledge about the area/domain which a given implementation of IBM IOC refers to and thus models of processes and organizations remain isolated from the specifics of the problem described by them. This is a serious limitation, since it prevents the full integration of knowledge about the implementation. The knowledge about the area/domain/problem in the above-mentioned approach is the knowledge that needs to be stored/processed in external systems.

In the next section we indicate an extension method of knowledge representation used in IBM IOC at the ontology level and we will identify tools for its integration with the existing solution.

3   Ontologies and a Controlled Natural Language in the Design Process of Smart Cities Systems

In the previous section we discussed the current market approach to designing the implementation of the Smart City based on IBM IOC. We have also pointed to an important limitation of the above approach which prevents the full integration of knowledge about the solution with the knowledge of the area/domain/problem the solution refers to. In this section, we show that the integration of semantic methods with a two-level IBM IOC configuration process, can fill this gap. Thus, we will create another level of the integration bus configuration - the ontology level. The section will be concluded with an introduction to modern methods of expressing knowledge in the form of (controlled) natural language, and we will describe a tool for integrating the ontology level with the other two levels of implementation of the IBM IOC integration bus. The resulting extension will become the tool basis for the Smart City pattern language described in the next section - a language allowing for a component-based decomposition of the combinatorial space of possible implementation solutions of the integration bus.

In the previous section we spoke of the MDA architecture called MOF which comes with IBM IOC, and we defined it as a four-layer architecture of meta-models. As shown in the figure (Fig. 4), the metameta-model referred to as M3 is the most general description of the world of models in MOF, it is the definition of the basic concepts of MOF. The M2 layer is a model of modeling languages such as UML or BPMN. M1 refers to models stored in the above-mentioned modeling languages, while M0 is the model proper that relates directly to the modeled world.

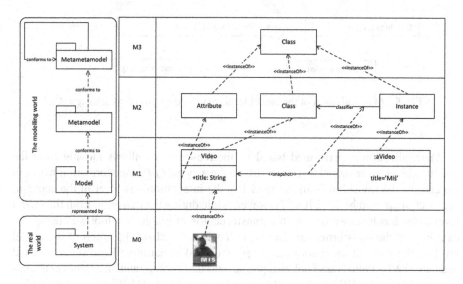

Fig. 4. General Meta-Object Facility architecture

The MOF M2 layer that defines the modeling languages (and concepts such as class, attribute, relationship or process) and the M3 layer that defines the most abstract concepts (such as the notion of being a subclass, the notion of being a relationship or being a specific part of the class/set - an instance) appear to be common also for semantic technologies from the W3C Semantic Initiative family. OWL [8] is integrated with MOF through the assignment of UML elements (M2 layer) to the corresponding OWL elements and the direct indication of the M3 layer representation in the basic OWL concepts.

It is thanks to the existence of the appropriate concepts between MOF and OWL that it is possible to extend the two-level IBM IOC configuration process (based on MOF) with the OWL semantic methods - methods which enable the recording of knowledge in the form of an ontology. Ontologies are tools to record knowledge about the area/domain/problem described by the implementation of the IOC integration bus. Moreover, in this way (having the above extension) we also extend our range of tools (until now they were the graph transformation languages) with systems for proving theorems (reasoners) operating in the formalism of the *SRQIQ* narrative logic [7, 9]. An official responsible for the implementation of a given solution, with the help of the reasoner, may find the confirmation that a given solution is useful in the context of a given area/domain/problem.

By applying a combination of MOF and OWL in the implementation of the IOC IBM bus, we add another level of the IOC bus implementation - the ontology level (O), while the implementation is extended with the model materialization stage on the basis of the ontology (Fig. 5).

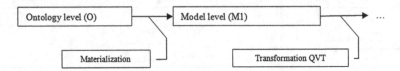

Fig. 5. Materialization of the model level (M1) on the basis of an ontology (O)

Materialization is performed based on the reasoner and allows moving from the ontology level to the model level similarly to the way the QVT transformation works in the case of the transition from the model to the integration bus level in the standard approach supported by IBM IOC. However, while during the transition from the model level to the bus level we deal with a transformation of graphs, the tool we created for carrying out the transformation (OntoTransTool) materializes the graph of the model level on the basis of the theory (ontology) recorded in narrative logic (of the formal database OWL). OntoTransTool, as input, accepts the ontology and on its basis generates an input file for the IBM Business Modeler (see Fig. 6) - (IBM Business Modeler is a tool, mentioned in the previous section, which is used to model the IBM IOC implementation from the level of the model - see Fig. 2).

```
C:\Windows\System32\cmd.exe                                                  _ □ ×

d:\OntoTransTool>OntoTransTool.exe --input "SmartCityOntology.encnl" --output "SmartCityModel.xml"
Engine setup...
Loading ontology into reasoner...
Processing Organizations...
Processing Activities...
Setting Up Ownership...
Creating IBM Business Modeller XML File...
Saveing...
...Done

d:\OntoTransTool>_
```

Fig. 6. Materialization of the M2 model from the ontology level (O) with the use of OntoTransTool

The next step after the integration of the ontology level with the two-level implementation process of the IOC integration bus is to select a tool for modeling ontology this at level. The best known tool for modeling ontology is Protégé, developed by Stanford University [13]. Protégé enables the creation and debugging of an ontology. It is equipped with a graphical user interface that allows the editing of an ontology in the interaction mode. We tried to use this tool as a tool for modeling at the ontology level (see Fig. 7). Unfortunately, its user requires the appropriate training. To understand the knowledge the user needs to learn the knowledge engineering methods used in the tool - which, unfortunately, in practice prevents the use of this tool in conducting dialogue with people interested in the implementation of the Smart City solution (they are not usually interested in the aspects of engineering knowledge).

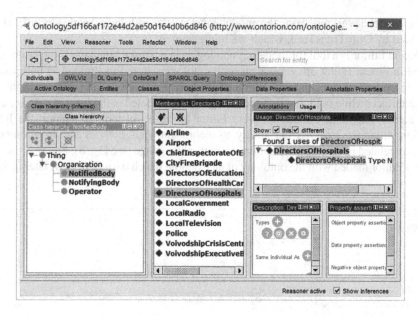

Fig. 7. Example configuration of the Smart City modeled on the ontology level (layer O) in Protégé

On the other hand, the opportunity to consult the ontology with interested parties is important for several reasons:

1. The client is involved in the configuration process - the client has a real impact not only on the requirements but also on the shape of the solution - which increases the chances of success
2. Ontology elements can be attached to the contract binding the person ordering a given IBM IOC implementation with the provider of the platform
3. The client has a chance to verify solutions (a priori) and thus has some evidence proving the correctness of decisions taken personally as early as at the design stage of implementation.

The above mentioned communication barrier regarding tool support is largely eliminated by so-called natural user interfaces - in particular, interfaces in the form similar to a natural language. This seems obvious (all people regardless of their field communicate in a natural language) and in the context of expressing an ontology, through the use of controlled natural languages (CNL) this was pointed out, among others, by Kuhn [11]; however, - as we will show in the section on implementation - often the natural language loses to graphical knowledge representation languages. In particular, graphical languages to record business processes such as BPMN - allowing for the representation of a process in the form of a graph - are easier to acquire for humans than their representation in a natural language. This graphical feature of languages recording business processes made us realize that it is worth reusing arti-facts that naturally arise at the level of the model during the modeling of an ontology. There was a need to develop bilateral transformation - both from the ontology level to the model level and from the model level to the ontology level which is also provided by OntoTransTool, which we implemented.

Fluent Editor is a tool for modeling ontologies in a natural language [3]. Ontologies stored in Fluent Editor are equivalent with ontologies stored in Protégé as both tools operate in the OWL technology, although in Protégé we are dealing with a graphical user interface, while in Fluent Editor ontologies are presented in the form of a document in a natural language. Including Fluent Editor in the set of tools to support the IOC configuration, we received the ability to record knowledge about the area/domain/ problem related to the implementation in a natural language (see Fig. 8). Figure 8 shows Fluent Editor loaded with the ontology of an example which we will discuss later in this paper.

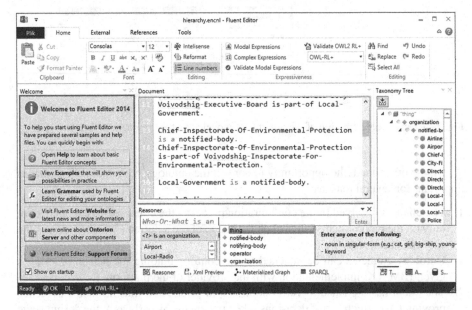

Fig. 8. Sample configuration of the Smart City modeled at the ontology level (layer O) in the Ontorion Fluent Editor tool.

To sum up: IBM IOC provides tools for the two-level integration bus implementation but ignores the aspect of integrating knowledge about the area/domain which a given implementation refers to and hence often (due to lack of tool support) models of processes and organizations remain in isolation from the specifics of the problems they describe. The third level we introduced - the ontology level - fills this gap. The modeling on the ontology level is performed with the use of controlled natural language, allowing for the direct use of ontologies saved in it to communicate with the client of the system. The integration of the ontology level is accomplished due to a tool we created, called OntoTransTool, which provides the possibility to materialize the ontology to the level of the model.

In the next section we introduce an ontological Smart City pattern language, based on the created tool database, which allows the decomposition of the combinatorial space of the possible implementation solutions of the integration bus and thus enables effective reuse of best design practices emerging during the work of experts.

4 Implementation of the Pattern Language for the Smart City

In the previous section we introduced the ontology level to the two-level implementation process of the IBM IOC integration bus. The ontology level allows the expression of knowledge about the area/domain/problem which a given implementation of the IBM IOC integration bus refers to in a manner consistent with the other two levels. The question arises - how to construct ontologies at this level in an optimal way. In the introduction, we set out to strive and create a tool which will enable the reuse of valuable expert knowledge regarding implementations. The best tool for this purpose, as proven by their growing popularity, are pattern languages.

Christopher Alexander [1] noted that cities are based on patterns. Currently, design patterns are a commonly used structure for exchanging solutions to recurrent problems. They are present in literature concerning urban planning (where they originated), information technology, pedagogy and many others. Alexander also notes that cataloging patterns creates a new value in the form of a language of patterns. A language, the words of which are names of individual patterns. At the same time, the pattern language allowing the resolution of problems which are combinations of simpler problems described by patterns-words, becomes a pattern: a pattern-language. A pattern is defined today as a method of documenting a solution of a recurrent design problem within a specific domain. In 1995, Gamma, Helm, Johnson and Vlissides [5] published the first directory of software design patterns, applying the ideas of Alexander to the field of software engineering.

One of the first attempts to formalize pattern languages can be found in a publication by Meszaros [12], in which we can find a self-definition of pattern language, as a pattern for creating more complicated patterns. This publication has become an inspiration to us and as such has given us a strong foundation to create a formal meta-language of patterns. We undertook to transpose the "patterns of writing patterns" described in the work of Meszaros onto the formalism of saving an ontology. The vehicle which allowed us to do so is controlled natural language.

It is worth focusing here on the language for recording an ontology - the controlled language which takes the form of a text file in Fluent Editor. Individual assertions (axioms) are separated with a period (.) and can be divided into a few basic groups:

1. Concept subsumption represents all cases where there is a need to specify (or constrain) the fact about a specific concept or instance (or expressions that evaluate the concept or instance) in the form of subsumption (e.g.: Every cat is a mammal, Pawel has two legs or One cat that is brown has red eyes).
2. Role (possibly complex) inclusion specifies the properties and relationships between roles in terms of the expressiveness of SROIQ(D) (e.g.: If X loves something that covers Y then X loves-cover-of Y).

3. Complex rules; If [body] then [head] expressions that are restricted to the DL-Safe SWRL subset [Fig. 1] of rules (e.g.: If a scrum-master is-mapped-to a provider and the scrum-master has-streamlining-assessment-processes-sprints-level equal-to 2 then the provider has-service-delivery-level equal-to 1 and the provider has-support-services-level equal-to 2).

4. Complex OWL expressions; the grammar allows the use of parentheses that can be nested if needed in the form of (that) e.g.: Every human is something (that is a man or a woman or a hermaphrodite).

5. Modal Expressions allow the expression of "knowledge about the knowledge" with CNL. This is to enforce fulfilling several properties of the knowledge. There are six modal words that can be used: must, should, can, must-not, should-not, can-not. All instances that are subject to validation against modal expressions are highlighted in Fluent Editor. Green means all requirements are fulfilled. Red means there is some requirement not met. Yellow means a warning. It appears when requirements with the "should" expression are not fulfilled. Light yellow means there is nothing wrong but it is marked just for informing the users. For example, regarding the statement "Every man can have a wife." - there is nothing wrong if there is a man that does not have a wife. For each modal expression different colors will appear as below:

   I.   X must Y (e.g. Every X must also Y) – if not, it will be marked in red.

  II.   X should Y (e.g. Every X should also Y) – if not, it will be marked in yellow.

 III.   X can Y (e.g. Every X can also Y) – if not, it will be marked in light yellow.

 IV.   X can-not Y (e.g. Every X cannot also Y) – if it is, then it will be marked in red.

  V.   X should-not Y (e.g. Every X should not also Y) – if it is, then it will be marked in yellow.

 VI.   X must-not Y (e.g. Every X must not also Y) – if it is, then it will be marked in light yellow.

The pattern language having the form of an ontology becomes a tool for managing trust towards the considered solution which is stored in it. It allows the application of computer methods as an important participant, which eliminates many errors generated by the "human factor." The ontology of the design pattern language should be self-describing, thus we shall start with the definition of basic concepts. The following example illustrates the use of a controlled language on the basis of a definition in the form of a pattern ontology "pattern" as well as a pattern "pattern language." The target structure of the ontology of the design pattern language is schematically drawn in Fig. 9. The pattern is defined here as a way to represent knowledge about solving the problem, it works in a certain context and resolves certain strengths prioritized by the above context. The pattern language meets the definition of a pattern (it is a pattern), solving a complex issue through a network of interrelated patterns.

Let us move on to the detailed record of the ontology structure of the pattern language, which is schematically illustrated in Fig. 9. Since we rely here on the taxonomy introduced by Meszaros, direct quotations from [12] are placed within single quotation marks. Please note the self-descriptive nature of the description of the structure in the controlled natural language.

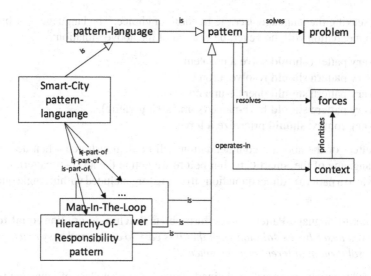

Fig. 9. Ontology structure of the design pattern language of the Smart City

§1. **The** problem **that** has-description **equal-to** *'How do you share a recurring solution to a problem with others so that it may be reused?'* **is** solved **by** Pattern-Pattern.

§2. **Every** pattern solves **at-most one** problem.

It is worth noting that the element describing the pattern-pattern solution requires the use (after Meszaros) of another pattern: the pattern of mandatory elements - here also direct quotations from [12] are included within single quotation marks.

§3. **The** problem **that** has-description **equal-to** *'How do you make sure **that** all necessary information is covered in a pattern?'* **is** solved **by** Mandatory-Elements-Present-Pattern.

The requirement as to the pattern format (which includes context, strength, solution and problem) manifests itself in the line below, in which we indicate by means of a modal expression that each model must carry out mandatory elements.

§4. **Every** pattern **must** realize Mandatory-Elements.

In turn, we point to the solution regarding the pattern of mandatory elements - indicating explicitly which elements must be included in the design pattern for it to be the pattern of mandatory elements.

§5. **If a** pattern solves **a** problem **and the** pattern resolves **a** force **and the** pattern has-name **(some string value) and** the pattern operates-in **a** context **and the** context prioritizes **the** force **then the** pattern realizes Mandatory-Elements.

We shall write down the solution offered by the pattern-pattern using the modal structure "should". The limitation it poses onto knowledge is ambiguous (sic) - in other words: we accept - as the creators of the specification - the possibility of breaking the contract, leaving the language user the freedom of choice. The language user breaking our recommendation will, however, be warned of the above violation:

§6. **Every** pattern **should** solve **a** problem.

§7. **Every** pattern **should** resolve **a** force.

§8. **Every** pattern **should** operate-in **a** context.

§9. **Every** pattern **should** have-name (**some string value**).

§10. **Every** context **should** prioritize **a** force.

The effect of the above recommendation will be demonstrated when defining the pattern language of the Smart City, but before we define the abstract pattern language (which is also a pattern) - direct quotations from [12] are included within single quotation marks.

§11. Pattern-Language-Pattern solves **the** problem **that** has-description **equal-to** *'How do you describe the solution such that it is easy to digest and easy to use parts of the solution in different circumstances?'*.

Pattern "language-patterns" as defined above generate a class of language patterns defined below, at the same time we require (it is mandatory) that it implements the pattern "language-pattern":

§12. **Every** pattern-language **is a** pattern.

§13. **Every** pattern-language **must** implement Pattern-Language-Pattern.

We require/ask (it is not mandatory) for the pattern language to be equipped with a syntactic layer manifested through the dictionary:

§14. **Every** pattern-language **should** use **a** pattern-dictionary.

And so it solves (again it is not mandatory) a complex problem. In addition, we define what we mean by the term "a complex problem":

§15. **Every** pattern-language **must** solve **a** complex-problem.

§16. **If a** problem(1) is-part-of **a** problem(2) **then the** problem(2) **is a** complex-problem.

And we indicate that compliance with our requests will result in the fulfillment of the mandatory requirement.

§17. **If a** pattern-language uses **a** pattern-dictionary **and the** pattern-language solves **a** complex-problem **then the** pattern-language implements Pattern-Language-Pattern.

Having defined abstract concepts, which are on a high-level conceptual basis, we can finally move on to the definition of a specific pattern language that will be adapted to the needs of the ontology level of the design and implementation process of the IBM IOC integration bus. And thus we define in the controlled language as follows:

§18. Smart-City-Pattern-Language **is a** pattern-language **and** operates-in Context-Of-Smart-City-Pattern-Language **and** has-name **equal-to** *'Smart City Pattern Language'*.

§19. Context-Of-Smart-City-Pattern-Language **is a** context **and** has-description **equal-to** *'A single pattern is insufficient to deal with all Smart City problems at hand.'*.

§20. Smart-City-Pattern-Language solves **the** problem **that** has-description **equal-to** *'How do you describe the solution for Smart City such that it is easy to digest and easy to use parts of the solution in areas of: a)People-First b)Business-Attractive c)Green d)Cheap (PBGC)'*.

§21. **The** force **that** has-description **equal-to** *'A single large Smart City solution may be too specific to the circumstance and impossible to reuse in other circumstances.'* **is** resolved **by** Smart-City-Pattern-Language **and is** prioritized **by** Context-Of-Smart-City-Pattern-Language.

§22. **The** force **that** has-description **equal-to** *'A complex Smart City solution may be hard to describe in a single pattern. A "divide and conquer" approach may be necessary to make the solution tractable.'* **is** resolved **by** Smart-City-Pattern-Language **and is** prioritized **by** Context-Of-Smart-City-Pattern-Language.

§23. **The** force **that** has-description **equal-to** *'Factoring the Smart City solution into a set of reusable steps can be very difficult. Once factored, the resulting pieces may depend on one another to make any sense.'* **is** resolved **by** Smart-City-Pattern-Language **and is** prioritized **by** Context-Of-Smart-City-Pattern-Language.

§24. **The** force **that** has-description **equal-to** *'Other pattern languages may want to refer to parts of the Smart City solution; they require some sort of "handle" for each of the parts to be referenced.'* **is** resolved **by** Smart-City-Pattern-Language **and is** prioritized **by** Context-Of-Smart-City-Pattern-Language.

§25. Smart-City-Pattern-Dictionary **is a** pattern-dictionary.

§26. Smart-City-Pattern-Language uses Smart-City-Pattern-Dictionary.

Above we defined the notion of the Smart City pattern language. The Smart City pattern language is a pattern that allows complex problems to be solved, which cannot be solved with a single pattern. The pattern language solves the problem of implementing a complex Smart City solution. The knowledge engineer operating in the ontology layer has the ability to use the Smart City pattern language, with concepts representing particular patterns. He is "protected" from its misuse by modal expressions (mandatory orders, requests or suggestions) included in the "pattern metalanguage".

To illustrate the work of the pattern language in practice, we shall define two patterns, and we will use them in an exemplary implementation. These will be patterns which occur commonly and naturally in all the cities, responding to real, common problems.

(1) Hierarchy of responsibility
(2) Man In The Loop-Resource conflict solver.

Of course these are only two from a whole spectrum of design patterns. So far we have managed to catalog ten patterns, however, a detailed description of all of the cataloged patterns goes beyond the scope of this paper. We shall list them in name only:

(3) Agregate KPIs,
(4) Mapping the KPIs,

(5) Predicting KPIs,
(6) Administrative SOP,
(7) Smart-Monitoring,
(8) Vectorization of streams,
(9) Social monitoring channel,
(10) Dynamic road sign.

Two patterns which will be described in detail were selected because of their subsequent use in the validation of the pattern language in a sample real implementation of the IBM IOC configuration bus.

4.1 Smart City Pattern: Hierarchy of Responsibility

The first pattern to be described here solves a key problem from the point of view of the city - the problem of responsibility for decisions. In this section we will show how the Smart City pattern language should be used for the purpose of expressing knowledge about a known and commonly acceptable solution to a given problem, in the form of a pattern. We operate at the ontology level of the extended configuration process of the IBM IOC integration bus.

The city is managed by the relevant authorities. The most common approach to city management is a hierarchy of responsibility. By establishing rules of communication flow, the hierarchy improves delegating tasks. Below there is a "hierarchy of responsibility" pattern. It is a part of the Smart City dictionary and it is also a problem that is solved by the Smart City Pattern Language Patterns:

§27. Context-Of-Hierarchy-Of-Responsibility-Pattern **is a** context **and** has-description **equal-to** *'Bez hierarchii przepływu oraz możliwości eskalacji problemu jest trudno zarządzać miastem'*.

§28. Hierarchy-Of-Responsibility-Pattern **is a** pattern **and** operates-in Context-Of-Hierarchy-Of-Responsibility-Pattern **and** has-name **equal-to** *'Hierarchy Of Responsibility'* **and** solves **the** problem **that** has-description **equal-to** *'Jak ułożyć hierarchię przepływu decyzji oraz eskalacji problemów?'* **and** is-part-of Smart-City-Pattern-Dictionary.

§29. **The** problem **that is** solved **by** Hierarchy-Of-Responsibility-Pattern is-part-of **the** problem **that is** solved **by** Smart-City-Pattern-Language.

§30. **The** force **that** has-description **equal-to** *'często kompetencje urzędow są zduplikowane lub niejednoznacznie określone'* **is** resolved **by** Hierarchy-Of-Responsibility-Pattern **and is** prioritized **by** Context-Of-Hierarchy-Of-Responsibility-Pattern.

The solution to this pattern requires defining a few truths. And so:

§31. **Every-single-thing that** manages **is a** management.
§32. **Every** city **must be** managed **by**.
§33. **Every** smart-city **is a** city.
§34. **Every** smart-management **is a** management.

We require that a Smart City is managed by a smart-management.

§35. **Every** smart-city **must be** managed **by a** smart-management.

§36. **Every** management **that** manages **a** smart-city **must be a** smart-management.

In addition, we show that the use of this pattern will result in the realization of the mandatory requirement.

§37. **If a** management implements Hierarchy-Of-Responsibility-Pattern **then the** management **is a** smart-management.

This sentence above (see §37) seems to be an oversimplification. Here, in order to validate a mandatory part correctly it is enough to declare its implementation (e.g. by stating: Management-X implements Hierarchy-Of-Responsibility-Pattern). However, the person who enters such a sentence becomes responsible for the explanation of how they managed to do it. Therefore it is also required that:

§38. **Every** management **that** implements Hierarchy-Of-Responsibility-Pattern **must** provide-implementation-details **(some** ontology-reference **value)**.

This forces the language user to justify the implementation of a given pattern. Going further, we can expand our requirements regarding the conditions under which we agree to the pattern being deemed as fulfilled: e.g. the verification of certain properties of the pattern that it must meet. In this paper, however, we keep the current approach, which is the simplest.

Let us add that a city managed using the pattern "Hierarchy-Of-Responsibility" is a city where there is an escalation of problems. It will be important in view of the next model, which requires that such an escalation is possible:

§39. **If a** management implements Hierarchy-Of-Responsibility-Pattern **then the** management **is a** smart-management **and the** management **is an** escalator-based-management.

The above example shows how patterns become interlaced in the pattern language. Patterns are not dependent on each other directly, but conceptually - they are intertwined at the level of concepts. As a result, some patterns "fit" to other patterns while others do not, and using the pattern language (its pragmatics) looks like matching suitable blocks from the set.

4.2 Smart City Pattern: Man-in-the-Loop-Resource-Resolver

Another significant problem that every city must face is the problem of limited resources. In this section we will present a solution to this recurrent problem, which is most commonly implemented. We also operate at the ontology level of the extended configuration process of the IBM IOC integration bus.

Conflicts arising during emergencies are associated with attempts to use limited resources in the context of the temporal coincidence of many emergencies. In particular, the interweaving of the same resources among equal domains can (in the event of a conflict) lead to a situation requiring the "best possible" decision to be taken swiftly.

The Man-In-The-Loop-Resource-Resolver pattern solves this problem by inserting a person into the decision-making process, who must assume responsibility for the decision. The person, however, is not alone, since the pattern finds the existence of an environment to raise escalation mandatory - such an environment is for example created by the Hierarchy of Responsibility pattern discussed earlier. This pattern is a part of the Smart City dictionary that we created and is a problem that is solved by the Smart City Pattern Language Patterns:

§40. Context-Of-Man-In-The-Loop-Resource-Resolver-Pattern **is a** context **and** has-description **equal-to** *'Smart-City wymaga zamodelowania kilku domen, wszystkie ww. domeny współdzielą ograniczone zasoby'*.

§41. Man-In-The-Loop-Resource-Resolver-Pattern **is a** pattern **and** operates-in Context-Of-Man-In-The-Loop-Resource-Resolver-Pattern **and** has-name **equal-to** *'Man In The Loop Resource Resolver'* **and** solves **the** problem **that** has-description **equal-to** *'W jaki sposób współdzielić ograniczne zasoby między różne domeny?'* **and** is-part-of Smart-City-Pattern-Dictionary.

§42. **The** problem **that is** solved **by** Man-In-The-Loop-Resource-Resolver-Pattern is-part-of **the** problem **that is** solved **by** Smart-City-Pattern-Language.

§43. **The** force **that** has-description **equal-to** *'Efektywne zarządzenie zasobami wymaga wiedzy eksperckiej'* **is** resolved **by** Man-In-The-Loop-Resource-Resolver-Pattern **and is** prioritized **by** Context-Of-Man-In-The-Loop-Resource-Resolver-Pattern.

§44. **The** force **that** has-description **equal-to** *'Niekiedy decyzje o przydziale zasobów wymagają zaangażowania wielu osób'* **is** resolved **by** Man-In-The-Loop-Resource-Resolver-Pattern **and is** prioritized **by** Context-Of-Man-In-The-Loop-Resource-Resolver-Pattern.

Below we present the definition of a city fragile to shared resources:

§45. **Every-single-thing that is** dealt-with **by something is a** domain.

§46. **Every-single-thing that is** used-resource **by something is a** resource.

§47. **If a** smart-city deals-with **a** domain(1) **and the** smart-city deals-with **a** domain(2) **and the** domain(1) **is-not-the-same-as the** domain(2) **and the** domain(1) uses-resource **a** resource(1) **and the** domain(2) uses-resource **a** resource(2) **and the** resource(1) **is-the-same-as the** resource(2) **then the** smart-city **is a** resource-fragile-smart-city.

It is required that the city management (which decides to deploy the above pattern) could escalate problems:

§48. **Every** management **that** manages **a** smart-city **that** implements Man-In-The-Loop-Resource-Resolver-Pattern **must be an** escalator-based-management.

§49. **Every** resource-fragile-smart-city **can** implement Man-In-The-Loop-Resource-Resolver-Pattern.

We define a condition the fulfillment of which will balance the shared city resources.

§50. **Every** resource-fragile-smart-city **must be a** resource-balanced-smart-city.

And we confirm the correctness of the solution provided by the above pattern. By implementing the above pattern, the city will be a balanced city.

§51. **Every** smart-city **that** implements Man-In-The-Loop-Resource-Resolver-Pattern **is a** resource-balanced-smart-city.

And as it was for the "Hierarchy of Responsibility" pattern, we require the details of the implementation of the solution to be provided:

§52. **Every** smart-city **that** implements Man-In-The-Loop-Resource-Resolver-Pattern **must** provide-implementation-details (**some** ontology-reference **value**).

To sum up, we presented the Smart City pattern language and presented two selected patterns in detail. These patterns have been selected due to their subsequent use in the validation of the pattern language. The validation will be conducted in the next section.

Examples of use of the Smart City pattern language.

The formalism presented above may be treated as a tool for selecting best practices-patterns from a directory. The key element of the pattern (which determines its existence) is the solution it has to offer. Patterns suggest these solutions and present their specification. The actual (real) implementation of a pattern must comply with the recommendation to be compatible with it. In other words: the art of using design patterns is the ability to select them, and - more importantly - the ability to materialize them - to fulfill them. Below there is an example of materializing the above-mentioned patterns.

We consider an example deployment of a Smart City in Gdansk. We declare as follows:

§53. Gdansk **is** managed **by** Gdansk-Government.
§54. Gdansk **is a** smart-city.

And by verifying the received knowledge – Gdańsk shimmers red (Fig. 10).

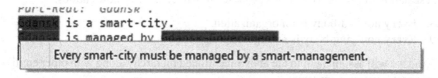

Fig. 10. Verification errors of the declared need for implementing the Smart City for the city of Gdansk

It turns out, in line with previous findings that: "*Every smart-city **must be** managed by a smart-management.*"

We know that:

§55. Gdansk-Government **is a** local-government.
§56. Gdansk-Government implements Hierarchy-Of-Responsibility-Pattern.

Re-validation indicates the need to justify the implementation of the pattern (Fig. 11).

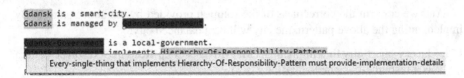

```
Gdansk is a smart-city.
Gdansk is managed by ████████████.

████████████ is a local-government.
████████████ implements Hierarchy-Of-Responsibility-Pattern
```
Every-single-thing that implements Hierarchy-Of-Responsibility-Pattern must provide-implementation-details

Fig. 11. Declaration of the implementation of the Smart City for the city of Gdansk after the information is provided about the type of pattern adopted for the management of the city

We will conduct the justification in a separate file describing the hierarchy and we will identify this file as a place where the aforementioned justification can be found. The first step is the correct location of the Smart-City management - in this case it is the Gdansk-Municipal-Crisis-Centre

§57.   Gdansk-Government **is** Gdansk-Municipal-Crisis-Centre[hier].

Going further:

§58.   Voivodship-Executive-Board is-part-of Local-Government.
§59.   Chief-Inspectorate-Of-Environmental-Protection is-part-of Voivodship
§60.   Inspectorate-For-Environmental-Protection.
§61.   Directors-Of-Health-Care-Facilities is-part-of The-"NZOZ".
§62.   Directors-Of-Health-Care-Facilities is-part-of Local-Government.
§63.   Directors-Of-Educational-Institutions is-part-of Local-Government.
§64.   Directors-Of-Care-Facilities is-part-of Local-Government.
§65.   Airport-Fire-Brigade is-part-of Airport.
§66.   City-Fire-Brigade is-part-of Provincial-Office.
§67.   Airport-Medical-Care is-part-of Airport.

The hierarchy defined in such a way also requires the definition of methods regarding the escalation of problems. For this purpose we will create two new entities:

§68.   **Every** notified-body **is an** organization.
§69.   **Every** notifying-body **is an** organization.

And we will assign them to the above-mentioned hierarchy along with other entities taking part in a given scenario:

§70.   Directors-Of-Hospitals **is a** notified-body.
§71.   Pilot **is a** notifying-body.
§72.   Airport **is a** notified-body.
§73.   Airline **is a** notified-body.
§74.   Police **is a** notified-body.
§75.   Directors-Of-Health-Care-Facilities **is a** notified-body.
§76.   Directors-Of-Educational-Institutions **is a** notified-body.
§77.   City-Fire-Brigade **is a** notified-body.
§78.   Voivodship-Crisis-Centre **is a** notified-body.
§79.   Voivodship-Executive-Board **is a** notified-body.
§80.   Chief-Inspectorate-Of-Environmental-Protection **is a** notified-body.

§81. Local-Government **is a** notified-body.
§82. Local-Radio **is a** notified-body.
§83. Local-Television **is a** notified-body.

This pattern implementation is contained in the described ontology "hierarchy.encnl". Typing in the above fact and subjecting the ontology to re-validation we get the green light. Currently, all mandatory requirements imposed on the city and its management are met, since only the city implements the "Hierarchy of Responsibility" pattern (Fig. 12).

```
Gdansk-Government is a local-government.
Gdansk-Government implements Hierarchy-Of-Responsibility-Pattern.

Gdansk-Government provides-implementation-details equal-to 'hierarchy.encnl'.
```

Fig. 12. Declaration of the implementation of the Smart City for the city of Gdansk after supplementing the information about the type of pattern adopted for the management of the city and indicating the means of its implementation

However, going further, it appears that in Gdansk we deal, among others, with an airport and the problem of air pollution and both of these domains share the need for the resource - Healthcare Facilities:

§84. Gdansk deals-with Airport-Domain.
§85. Gdansk deals-with Air-Pollution-Domain.
§86. Airport-Domain has-name **equal-to** *'Gdansk Airport Domain'*.
§87. Air-Pollution-Domain has-name **equal-to** *'Gdansk Air Pollution Domain'*.
§88. Airport-Domain uses-resource Healthcare-Facilities.
§89. Air-Pollution-Domain uses-resource Healthcare-Facilities.

And in result, Gdańsk becomes resource-fragile (Fig. 13).

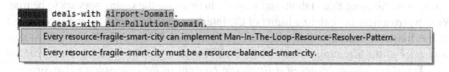

```
       deals-with Airport-Domain.
       deals-with Air-Pollution-Domain.
 Every resource-fragile-smart-city can implement Man-In-The-Loop-Resource-Resolver-Pattern.
 Every resource-fragile-smart-city must be a resource-balanced-smart-city.
```

Fig. 13. Further verification indicates that there is a resource conflict

The solution suggested by the validator is to implement the Man-In-The-Loop-Resource-Resolver-Pattern. The protocols of procedure arise from respective laws and are associated with respective domains. The protocols describe processes that are best presented in the processes modeling language. At this point, we return to the IBM IOC model level, as the modeled processes are at the level of M1 MOF - so it is best if it is supported by a tool for modeling in BPMN, such as the IBM Business Modeler. It appears that the process of IOC configuration is not a one-way process but a process operating in a feedback loop (Fig. 14).

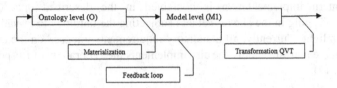

Fig. 14. Feedback in the modeling process arising due to the need for process modeling

Of course, this protocol is equivalent to the following (completely unintelligible) record in CNL:

§90.   Ev-000 **is a** start-event.

Act-001 **is a** activity-pm-10-measure-a.
Act-003 **is a** activity-pm-10-measure-b.
...
Act-019 **is a** activity-send-ambulances.
Ev-020 **is a** end-event.
Ev-021 **is a** start-event.
...
Ev-007 follows-if-false Act-001.
Act-005 follows-if-true Act-003.
Act-006 follows-if-false Act-003.
...
Ev-020 follows Act-019.
Act-022 follows Ev-021.
Act-023 follows Act-022.
...

Natural language shows its weakness here. A weakness illustrated in the saying "a picture is worth more than a thousand words". In this case the saying appears to be true. We can prove here the existing limits of the interface based on natural language - these are the limits of usability, limits that are rediscovered. Richard Riehle in [15] states: *"look back at the fascination in all we had with our flowcharting as a gigantic waste of time. Yet, there was a kernel of a linguistic concept in those templates that survives today."* Unfortunately diagrams have a major drawback - they represent networks - graphs - relationships between entities, while natural language sentences represent the workings of the world - they represent the truth that is realized (materialized) in graphs.
    Here are two diagrams:

1. The protocol of procedure in the case of excessive levels of PM10 (Fig. 15)

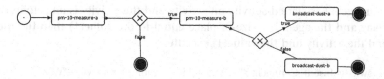

Fig. 15. The protocol of procedure in the case of excessive levels of PM10 in BPMN

2. The protocol of procedure in the case of an emergency landing of an aircraft depending on its size (Fig. 16)

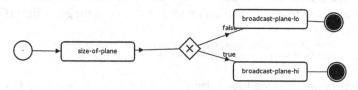

Fig. 16. The protocol of procedure in the case of emergency landing depending on the size of the aircraft

In constructing a "hierarchy of responsibility," we had to designate a "trail" of problems escalation. This trail is essential if we want to locate the person who can make decisions in case of a conflict of resources. We assume that the person does it rationally. Since, however, the person's decisions are rational, they can be substituted by automatic decisions. Proving this thesis requires the use of non-monotonic knowledge management systems (allowing for the removal of conclusions depending on the circumstances), and thus remaining beyond the scope of this paper. It is worth presenting examples of experimental rules written in the non-monotonic formalism. And thus, for our example:

§91. **If an** agent has-started-activity **an** activity **and the** activity **is an** activity-pm-10-measure-a **and the** agent has-pm-10-measurement **equal-to the value**(1) **then for the** agent **and the** activity **and the value**(1) **execute**

```
KnowledgeInsert("Comment:\""+agent + " is executing " + activity+"\".");
KnowledgeDelete(agent +" has-started-activity " + activity+".");
KnowledgeInsert(agent +" has-condition equal-to " + (value[1]>=50?"true":"false")+".");
KnowledgeInsert(agent +" has-finished-activity " + activity+".");
```

§92. **Every** activity-pm-10-measure-b **is an** activity.

§93. **If an** agent has-started-activity **an** activity **and the** activity **is an** activity-pm-10-measure-b **and the** agent has-pm-10-measurement **equal-to the value**(1) **then for the** agent **and the** activity **and the value**(1) **execute**

```
KnowledgeInsert("Comment:\""+agent + " is executing " + activity+"\".");
KnowledgeDelete(agent +" has-started-activity " + activity+".");
KnowledgeInsert(agent +" has-condition equal-to "+(value[1]>=300?"true":"false")+".");
KnowledgeInsert(agent +" has-finished-activity " + activity+".");
```

§94. **Every** activity-size-of-plane **is an** activity.

§95. **If an** agent has-started-activity **an** activity **and the** activity **is an** activity-size-of-plane **and the** agent has-size-of-plane **equal-to the value**(1) **then for the** agent **and the** activity **and the value**(1) **execute**

```
KnowledgeInsert("Comment:\""+agent + " is executing " + activity+"\".");
KnowledgeDelete(agent +" has-started-activity " + activity+".");
KnowledgeInsert(agent +" has-condition equal-to "+(value[1]>=150?"true":"false")+".");
KnowledgeInsert(agent +" has-finished-activity " + activity+".");
```

Concluding our discussion and saving the above solution in the 'man-in-the-loop.encnl' file and declaring in our ontology the above solution as an explanation of the pattern implementation, the validator once more gives us the green light (Fig. 17).

```
Gdansk implements Man-In-The-Loop-Resource-Resolver-Pattern.
Gdansk provides-explanation equal-to 'man-in-the-loop.encnl'.
```

Fig. 17. Declaration of the implementation of the Smart City for the city of Gdańsk after supplying information about the type of pattern adopted for resolving the resource conflict and indicating the means of its implementation

To sum up: in this section we presented the validation of the previously introduced pattern language. This validation involved the creation of a configuration ontology of the IBM-IOC integration bus on the basis of the pattern language. A full cycle of work with the pattern language was shown. We found that the cascade model of transition from the ontology (O) to the model level (M1) should be changed by adding a feedback loop, which allows the supplementation of the range of tools with an adequate transformation.

5   Criticism of Design Patterns

Criticism of the design patterns conducted, among others, by Peter Norwig[1] states that design patterns are solutions to problems generated by the language itself (in which they are stored or which they refer to). In other words - the pattern language tries to compensate for language problems of the language which it reaches - namely it solves the imperfections that the language itself should eliminate. Norwig operates in the world of software, but he is right to present most of the GOF patterns as "artificial problems" in languages such as LISP and Dylan which do not generate such problems. It seems, however, that Norwig, criticising the patterns of software, neglects the attributes of a pattern and identifies the pattern with the solution. However, the pattern remained and remains only a way to exchange experiences regarding solutions to recurring problems. From this point of view, the programming language is an important part of the pattern context and hence Norwig's criticism can be fended off by stating – the context of GOF patterns does not include languages such as LISP or Dylan.

---

[1] http://norvig.com/design-patterns/.

In our understanding, a pattern is a way of distributing experience. Its specific form, and the number of implementations of the idea of design patterns leads us to believe that the attributes of a pattern are a universal method. What is more - the Smart City pattern language which we introduced is in the form of formal ontologies which can be proved, though it remains only a means to exchange experiences about solutions to common problems. A means - as we are trying to show - worthy of further research.

6 Summary

Designing Smart Cities in its essence is an issue of urban planning as urban structures, the object of study of urban design, allow for the development of the concept of planning. The Smart City implementation project is the realization of a planning concept so it is in line with the aforementioned definition. The Smart City design patterns that we propose are analogous to design patterns in urban planning. In other words, the architecture of the old implementation of the Smart Cities integration bus, following the architecture of the city in which it is implemented, remains the architecture of an IT system. It is therefore a virtual extension of a given city. Applying the discovery of Christopher Alexander in the context of the implementation of the Smart City integration bus, we make a full circle - the patterns again refer to cities, this time, however, they operate on a virtual model/extension of the city.

In this paper, we presented the solution to problems associated with the reuse of extremely valuable knowledge that arises during the implementation of the Smart City integration bus. We proposed the use of a natural language matched to the issue, formally defined and equipped with adequate tools of the Smart City pattern language. What is important, this language gives a very powerful tool to authorities responsible for decisions - the ability to verify the solution by reasoners. The formal methods based on reasoners are the most powerful tool to build confidence in the proposed solution.

On the other hand, since making decisions - such as the decision to choose the configuration of the Smart City integration bus - can be supported by a computer, the question is: whether rational decisions can be made by a machine? Here, unfortunately, for us humans, the answer seems to be positive. The maximum understanding of knowledge "at one time" by humans seems to be indicated by the problem of "driving". Driving a car requires the knowledge of the Highway Code, the application of that knowledge as well as the ability to make quick operational decisions. On the other hand, as demonstrated by Nobel Prize winner Daniel Kahneman [10], analytical thinking costs people a lot of energy. Assimilation of the "whole" knowledge in a given domain by a human being reaches a certain limit determined by the time required for understanding the knowledge, and the full depth of understanding becomes, in a way, unattainable. Mathematicians can be proof of this, they reason deeply but mathematics itself (at least in comparison with the soft sciences) develops slowly. Theoretically, for computers the time to reach the depth of understanding is much shorter and the processing of knowledge seems to be limited only by the laws of logic. Here PCs gain an advantage over us and the idea of equipping Smart Cities with a solution, such as IBM Watson would seem to be interesting.

References

1. Alexander, C., Ishikawa, S., Silverstein, M.: A Pattern Language: Towns, Buildings, Construction (Center for Environmental Structure Series). Oxford University Press, Oxford (1977)
2. Bhowmick, A.: IBM Intelligent Operations Center for Smarter Cities Administration Guide. IBM Corporation, International Technical Support Organization (2012)
3. Cognitum: Fluent Editor 2014 - Ontology Editor (2014). http://www.cognitum.eu/semantics/FluentEditor/
4. Fowler, M.: Patterns of Enterprise Application Architecture (A Martin Fowler Signature Book). Addison-Wesley, Reading (2003)
5. Gamma, E., Helm, R., Johnson, R., Vlissides, J.: Design Patterns: Elements of Reusable Object-Oriented Software. Addison-Wesley Longman Publishing Co., Inc., Boston (1995)
6. Glimm, B., Horridge, M., Parsia, B., Patel-Schneider, P.: A syntax for rules in OWL In: Hoekstra, R., Patel-Schneider, P.F. (eds.) OWLED, vol. 529. CEUR Workshop Proceedings (2008). CEUR-WS.org
7. Goczyła, K.: Ontologie w systemach informatycznych. Akademicka Oficyna Wydawnicza EXIT (2011)
8. Hitzler, P., Krötzsch, M., Parsia, B., Patel-Schneider, P., Rudolph S.: OWL 2 web ontology language primer. In: W3C Recommendation, World Wide Web Consortium (2009)
9. Horrocks I., Kutz, O., Sattler, U.: The even more irresistible SROIQ. In: Doherty, P., Mylopoulos, J., Welty, C.A. (eds.) KR, pp. 57–67. AAAI Press (2006)
10. Kahneman, D.: Thinking, Fast and Slow. Farrar, Straus and Giroux, New York (2011)
11. Kuhn, T.: How to evaluate controlled natural languages. In: Fuchs, N.E. (ed.) Pre-Proceedings of the Workshop on Controlled Natural Language (CNL 2009), vol. 448. CEUR Workshop Proceedings (2009). CEUR-WS
12. Meszaros G., Doble, J.: A pattern language for pattern writing. In: Pattern languages of program design 3, pp. 529–574. Addison-Wesley Longman Publishing Co. (1997)
13. Musem, M, Noy N., Nyulas, C., O'Connor, M., Redmond, T., Tu, S., Tudorache, T., Vendetti, J. and Stanford School of Medicine: Protégé (2010). http://protege.stanford.edu
14. OMG: Business Process Model and Notation (BPMN), Version 2.0 (2011)
15. Riehle, R.: Linguistic continuity in software engineering. ACM SIGSOFT Softw. Eng. Notes 31(1), 1–5 (2006)
16. Rozenberg, G.: Handbook of Graph Grammars and Computing by Graph Transformation: Volume I Foundations. World Scientific Publishing Co., Inc., River Edge (1997)

Text Classification Using "Anti"-Bayesian Quantile Statistics-Based Classifiers

B. John Oommen[1(✉)], Richard Khoury[2], and Aron Schmidt[3]

[1] School of Computer Science, Carleton University, Ottawa K1S 5B6, Canada
oommen@scs.carleton.ca
[2] Department of Computer Science and Software Engineering,
Laval University, Quebec City G1V 0A6, Canada
richard.khoury@ift.ulaval.ca
[3] Department of Software Engineering, Lakehead University,
Thunder Bay P7B 5E1, Canada
aschmid1@lakeheadu.ca

Abstract. The problem of Text Classification (TC) has been studied for decades, and this problem is particularly interesting because the features are derived from *syntactic or semantic* indicators, while the classification, in and of itself, is based on *statistical* Pattern Recognition (PR) strategies. Thus, all the recorded TC schemes work using the fundamental paradigm that once the statistical features are inferred from the syntactic/semantic indicators, the classifiers themselves are the well-established ones such as the Bayesian, the Naïve Bayesian, the SVM etc. and those that are neural or fuzzy. In this paper, we shall demonstrate that by virtue of the skewed distributions of the features, one could advantageously work with information latent in certain "non-central" quantiles (i.e., those distant from the mean) of the distributions. We, indeed, demonstrate that such classifiers exist and are attainable, and show that the design and implementation of such schemes work with the recently-introduced paradigm of Quantile Statistics (QS)-based classifiers(The foundational properties for CMQS (for generic and some straightforward distributions) were initially described in [17]. Their properties for uni-dimensional distributions of the exponential family are included in [9], and for multi-dimensional distributions in [18]. The authors of [17], [9] and [18] had initially proposed their results as being based on the *Order*-Statistics of the distributions. This was later corrected in [19], where they showed that their results were rather based on their *Quantile* Statistics.). These classifiers, referred to as Classification by Moments of Quantile Statistics (CMQS), are essentially "Anti"-Bayesian in their *modus operandi*. To achieve our goal, in this paper we demonstrate the

The authors are grateful for the partial support provided by NSERC, the Natural Sciences and Engineering Research Council of Canada. A preliminary version of this paper was presented at ICCCI'15, the *2015 International Conference on Computational Collective Intelligence Technologies and Applications*, in Madrid, Spain, in September 2015. The paper was a *Plenary/Keynote* Talk at the conference. The first author is also an *Adjunct Professor* with the University of Agder in Grimstad, Norway.

© Springer-Verlag GmbH Germany 2016
N.T. Nguyen et al. (Eds.): TCCI XXV, LNCS 9990, pp. 101–126, 2016.
DOI: 10.1007/978-3-662-53580-6_7

power and potential of CMQS to describe the *very* high-dimensional TC-related vector spaces in terms of a limited number of "outlier-based" statistics. Thereafter, the PR task in classification invokes the CMQS classifier for the underlying multi-class problem by using a linear number of pair-wise CMQS-based classifiers. By a rigorous testing on the standard 20-Newsgroups corpus we show that CMQS-based TC attains accuracy that is comparable to the best-reported classifiers. We also propose the potential of fusing the results of a CMQS-based methodology with those obtained from a more traditional scheme.

Keywords: Text Classification · Quantile Statistics (QS) · Moments of QS · Classification by the Moments of Quantile Statistics (CMQS) · Prototype Reduction Schemes

1 Introduction

Text Classification (TC) is the challenge of associating a given unknown text document with a category selected from a predefined set of categories (or classes) based on its content. This problem has been studied since the 1960's [16], but it has taken a special importance in recent years as the sheer amount of text available has increased super-exponentially – thanks to the internet, text-based communications such as e-mail, tweets and text messages, and the numerous book-digitization projects that have been undertaken by the various publishing houses. Over the decades, many approaches have been proposed to accomplish this goal. When it concerns classification and Pattern Recognition (PR), the TC problem is particularly interesting both from an academic and a research perspective. This is because, whereas the features in TC are derived from *syntactic or semantic* indicators, the classification, in and of itself, is based on *statistical*, neural or fuzzy strategies.

Statistical PR is the process by which unknown *statistical* feature vectors are categorized into groups or classes based on their *statistical* components [3]. The field of statistical PR has been so well developed that it is not necessary for us to survey the field here. Suffice it to mention that all the recorded TC schemes work using the fundamental paradigm that once the statistical features are inferred from the *syntactic or semantic* indicators, the classifiers themselves are the well-established *statistical*, neural or fuzzy ones such as the Bayesian, Naïve Bayesian, Linear Discriminant, the SVM, the Back-propagation etc.

The goal of this paper is to show that we can achieve TC using "Anti"-Bayesian quantile statistics-based classifiers which only use information contained in, let us say, non-central quantiles (which are sometimes outliers) of the distributions, and also achieve this task by operating with a philosophy that is totally contrary to the acclaimed Bayesian paradigm. Indeed, the fact that such a classification can be achieved is, strictly speaking, not easy to fathom.

1.1 Motivation for the Paper

To motivate this paper and to place its contribution the right context, we present the following simple example. Consider the problem of distinguishing a document that belongs to one of two classes, namely, *Sports* or *Business*. It is obvious that one can trivially distinguish them if we merely considered those words which occurred frequently in one class and not the other, for example, "football" and "basketball" *versus* "dollars" and "euros". Our hypothesis is that it is not *merely* these truly "distinguishing" words that possess "discriminating" capabilities. We intend to demonstrate that there are "outliers" quantiles of the words which occur in both categories, and which also can be used to achieve the classification. Hopefully, this would be both a pioneering and remarkable result.

It should, first of all, be highlighted that we do not intend to obtain a classification that *surpasses* the behavior of the scheme that involves a Bayesian strategy invoking the truly "distinguishing" words. Attempting to do this would be tantamount to accomplishing the impossible, because the Bayesian approach maximizes the *a posteriori* probability and it thus yields the optimal hallmark classifier. What we endeavor to do is to show that if we use the above-mentioned non-central quantiles and work within an "Anti"-Bayesian paradigm using only *these* quantile statistics, we can obtain accuracies comparable to this optimal hallmark! Indeed, we demonstrate that a near-optimal solution can be obtained by invoking counter-intuitive features *when they are coupled with* a counter-intuitive PR paradigm.

As a backdrop, we note that the basic concept of traditional *parametric* classification is to model the classes based on the assumptions related to the underlying class *distributions*, and this has been historically accomplished by performing a learning phase in which the moments, i.e., the mean, variance etc. of the respective classes are evaluated. However, there have been some families of indicators (or distinguishing quantifiers) that were until recently, noticeably, *uninvestigated* in the PR literature. Specifically, we refer to the use of phenomena that have utilized the properties of the *Quantile Statistics* (QS) of the distributions. This has led to the "Anti"-Bayesian methodology alluded to.

It is expedient to examine how these two fields, namely those of *statistical* and *syntactic* PR are "merged". Before we embark on this, we shall briefly describe some preliminary concepts used in TC and in fields that are related.

1.2 Preliminaries: Documents, Terms, BOWs and Similarity Measurements

Detecting textual similarities is an important building block in the structuring (for example, clustering) of collections of documents, in Information Retrieval (IR), and in classification. The art relies on the computation of indices quantifying textual similarities, and on measuring the distance between a given query and documents, or the similarity between multiple documents. Detecting the relevance of a document to a specific user's query is a highly pertinent problem. Ranking documents is also a task that can be done to prune a large collection of documents before presenting them to the user. To perform such actions, the system needs a metric to quantify the similarity/dissimilarity between the documents. Furthermore, in order to be able to apply good measures, the documents must also be represented in a suitable model or structure. One of the most commonly used models is the Vector Space Model (VSM) explained below.

The Vector Space Model: The VSM, (also called the vector model) was first presented by Salton *et al.* [13] in 1975, and used as a part of the SMART[1] Information Retrieval System developed at Cornell University. The model involves an algebraic system for document representation, where, in the processing of the text, the model uses vectors of identifiers, where each identifier is normally a term or a token. For the purpose of the representation of documents, the VSM would be a list of vectors for all the terms (words) that occur in the document. Since a document can be viewed as a long string, each term in the string is given a correlating value, called a weight. Each vector consists of the identifier and its weight. If a certain term exists in the document, the weight associated with the term is a non-zero value, commonly a real number in the interval $[0, 1]$. The number of terms represented in the VSM is determined by the vocabulary of the corpus.

Although the VSM is a powerful tool in document representation, it has certain limitations. The obvious weakness is that it requires vast computational resources. Also, when adding new terms to the term space, each vector has to be recalculated. Another limitation is that "long" documents are not represented optimally with regard to their similarity values as they lead to problems related to small scalar products and large dimensionalities. Furthermore, the model is sensitive to semantic content, for example, documents with similar content but different term vocabularies will not be associated, which is, really, a false negative match. Another important limitation that is worth mentioning is that search terms must match the terms found in the documents precisely, because substrings might result in a false positive match. Last, but not least, this model does not preserve the order in which the terms occur in the document. Despite these limitations, the model is useful, and can be improved in several ways, but details of these improvements are omitted here.

A text classification algorithm, typically, begins with a representation involving such a collection of terms, referred to as the Bag-of-Words (BOW) representation [16]. In this approach, a text document D is represented by a vector $[w_0, w_1, \ldots, w_{N-1}]$, where w_i is the occurrence frequency of word i in that document. This, so-called, "word" vector is then compared to a representation of each category, to find the most similar one. A straightforward way of implementing this comparison is to use a pre-computed BOW representation of each category from a set of previously-available representative documents used for the training of the classifier, and to compute for example, a similarity between the vector associated with each category and the vector associated with the document to be classified. The cosine similarity measure is just one of a number of "metrics" that can be used to achieve the comparison. More refined methods replace simple word counts with weights that take into account the typical occurrence frequencies of words across

---

[1] SMART is an abbreviation for Salton's Magic Automatic Retriever of Text.

categories, in order to reduce the significance imparted to common words and to enhance domain-specific ones.

Salton also presented a theory of "term importance" in automatic text analysis in [14]. There, he stated that the terms which have value to a document are those that highlight differences or contrasts among the documents in the corpus. He noted that: "*A single term can decrease the document similarity among document pairs if its frequency in a large fraction of the corpus is highly variable or uneven.*" One very simple term weighting scheme is the so-called Term Count Model, where the weight of each term is simply given by counting the number of occurrences (also called the set of Term Frequencies) of the term[2].

The TFIDF Scheme: The problem with a simplistic "frequency-based" scheme is that it is inadequate when it concerns the repetition of terms, and that it actually favors large documents over shorter documents. Large documents obtain a higher score merely because they are longer, and not because they are more relevant. The Term Frequency-Inverse Document Frequency (TFIDF) weighting scheme achieves what Salton described in his term importance theory by associating a weight with every token in the document based on both local information from individual documents and global information from the entire corpus of documents. The scheme assumes that the importance of a term is proportional to the number of documents that the term appears in. The TFIDF scheme models both the importance of the term with respect to the document, and with respect to the corpus as a whole [12,14]. Indeed, as explained in [15], the TFIDF scheme weights a term based on how many times it is represented in a document, and this weight is simultaneously negatively biased based on the number of documents it is found in. Such a weighting philosophy can be seen to have the effect that it correctly predicts that very common terms, occurring in a large number of documents in the corpus, are not good discriminators of relevance, which is what Salton required in his theory of term importance.

Although the formal expression for the TFIDF is also given in a later section, it is pertinent to mention that the TFIDF is computationally efficient due to the high degree of sparsity of most of the vectors involved, and by using an appropriate inverted data structure for an efficient representation mechanism. Indeed, it is considered to be a reasonable off-the-shelf metric for long strings and text documents[3]. Other alternatives, based on information gain and chi-squared metrics [2], have also been proposed.

---

[2] The formal definitions for the TF and the TFIDF are given in Sect. 4.3.

[3] Since the static TFIDF weighting scheme presented above becomes inefficient when the system has documents that are continuously arriving, for example, systems used for online detection, the literature also reports the use of the *Adaptive* TFIDF. The *Adaptive* IDF can be efficiently used for document retrieval after a sufficient number of "past" documents have been processed. The initial IDF values are calculated using a retrospective corpus of documents, and these IDF values are then updated incrementally. The literature also reports other metrics of comparison, such as the Jaccard similarity, but since this is not the primary concern of this paper, we will not elaborate on these here.

The question of how these statistical features (BOW frequency or TFIDF) are incorporated into a TC that also uses statistical PR principles is surveyed in more depth in Sect. 2.

1.3 Contributions of This Paper

The novel contributions of this paper are:

- We demonstrate that text and document classification can be achieved using an "Anti"-Bayesian methodology;
- To achieve this "Anti"-Bayesian PR, we show that we can utilize syntactic information that has not been used in the literature before, namely the information contained in the symmetric quantiles of the distributions, and which are traditionally considered to be "outlier"-based;
- The results of our "Anti"-Bayesian PR is not highly correlated with the results of any of the traditional TC schemes, implying that one can use it in conjunction with a traditional TC scheme for an ensemble-based classifier;
- Since the features and methodology proposed here are distinct from the state-of-the-art, we believe that a strategy that incorporates the fusion of these two distinct families has great potential. This is certainly an avenue for future research.

As in the case of the quantile-based PR results, to the best of our knowledge, the pioneering nature and novelty of these TC results hold true.

1.4 Paper Organization

The rest of the paper is organized as follows. First of all, in Sect. 2, we present a brief, but fairly comprehensive overview of what we shall call, "Traditional Text Classifiers". We proceed, in Sect. 3 to explain how we have adapted "Anti"-Bayesian classification principles to text classification, and follow it in Sects. 4 and 5 to explain, in detail, the features used, the datasets used, and the experimental results that we have obtained. A discussion of the results has also been included here. Section 6 concludes the paper, and presents the potential avenues for future work.

2   Background: Traditional Text Classifiers

Apart from the methods presented above, many authors have also looked at ways of enhancing the document and class representation by including not only words but also bigrams, trigrams, and n-grams in order to capture common multi-word expressions used in the text [4]. Likewise, character n-grams can be used to capture more subtle class distinctions, such as the distinctive styles of different authors for authorship classification [10]. While these approaches have, so far, considered ways to enrich the representation of the text in the word vector, other authors have attempted to augment the text itself by adding extra information into it, such as synonyms of the words taken from a thesaurus, be it a specialized custom-made one for a project such as the affective-word thesaurus built in [8], or, more commonly, the more general-purpose linguistic ontology, *WordNet* [5].

Adding another generalization step, it is increasingly common to enrich the text not only with synonymous words but also with synonymous *concepts*, taken from domain-specific ontologies [22] or from Wikipedia [1]. Meanwhile, in an opposing research direction, some authors prefer to simplify the text and its representation by reducing the number of words in the vectors, typically by grouping synonymous words together using a Latent Semantic Analysis (LSA) system [7] or by eliminating words that contribute little to differentiating classes as indicated by a Principal Component Analysis (PCA) [6]. Other authors have looked at improving classification by mathematically transforming the sparse and noisy category word space into a more dense and meaningful space. A popular approach in this family involves Singular Value Decomposition (SVD), a projection method in which the vectors of co-occurring words would project in similar orientations, while words that occur in different categories would be projected in different orientations [7]. This is often done before applying LSA or PCA modules to improve their accuracy. Likewise, authors can transform the word-count space to a probabilistic space that represents the likelihood of observing a word in a document of a given category. This is then used to build a probabilistic classifier, such as the popular Naïve-Bayes' classifier [11], to classify the text into the most probable category given the words it contains.

An underlying assumption shared by all the approaches presented above is that one can classify documents by comparing them to a representation of what an average or typical document of the category should look like. This is immediately evident with the BOW approach, where the category vector is built from average word counts obtained from a set of representative documents, and then compared to the set of representative documents of other categories to compute the corresponding similarity metric. Likewise, the probabilities in the Naïve-Bayes' classifier and other probability-based classifiers are built from a corpus of typical documents and represent a general rule for the category, with the underlying assumption that the more a specific document differs from this general rule, the less probable it is that it belongs to the category. The addition of information from a linguistic resource such as a thesaurus or an ontology is also based on this assumption, in two ways. First, the act itself is meant to add words and concepts that are missing from the specific document and thus make it more like a typical document of the category. Secondly, the development of these resources is meant to capture general-case rules of language and knowledge, such as "these words are typically used synonymously" or "these concepts are usually seen as being related to each other."

The method we propose in this paper is meant to break away from this assumption, and to explore the question of whether there is information usable for classification outside of the norm, at "the edges (or fringes) of the word distributions", which has been ignored, so far, in the literature.

3   CMQS-Based Text Classifiers

3.1   How Uni-dimensional "Anti"-Bayesian Classification Works

We shall first describe how uni-dimensional "Anti"-Bayesian classification works, and then proceed to explain how it can be applied to TC, which, by definition, involves PR in a highly multi-dimensional feature space[4].

Classification by the Moments of Quantile Statistics[5], (CMQS) is the PR paradigm which utilizes QS in a pioneering manner to achieve optimal (or near-optimal) accuracies for various classification problems[6]. Rather than work with "traditional" statistics (or even sufficient statistics), the authors of [17] showed that the set of *distant* quantile statistics of a distribution do, indeed, have discriminatory capabilities. Thus, as a *prima facie* case, they demonstrated how a generic classifier could be developed for any uni-dimensional distribution. Then, to be more specific, they designed the classification methodology for the Uniform distribution, using which the analogous classifiers for other symmetric distributions were subsequently created. The results obtained were for symmetric distributions[7], and the classification accuracy of the CMQS classifier exactly attained the optimal Bayes' bound. In cases where the symmetrtic QS values crossed each other, one invokes a *dual* classifier to attain the same accuracy.

Unlike the traditional methods used in PR, one must emphasize the fascinating aspect that CMQS is essentially "Anti"-Bayesian in its nature. Indeed, in CMQS, the classification is performed in a counter-intuitive manner i.e., by comparing the testing sample to a few samples *distant* from the mean, as opposed to the Bayesian approach in which comparisons are made, using the Euclidean or a Mahalonibis-like metric, to *central* points of the distributions. Thus, opposed to a Bayesian philosophy, in CMQS, the points against which the comparisons are made are located at the positions where the Cumulative Distribution Function (CDF) attains the percentile/quantile values of $\frac{2}{3}$ and $\frac{1}{3}$, or more generally, where the CDF attains the percentile/quantile values of $\frac{n-k+1}{n+1}$ and $\frac{k}{n+1}$.

In [9], the authors built on the results from [17] and considered various symmetric and *asymmetric* uni-dimensional distributions within the exponential family such as the Rayleigh, Gamma, and Beta distributions. They again proved that CMQS had an accuracy that attained the Bayes' bound for symmetric distributions, and that it was very close to the optimal for asymmetric distributions.

---

[4] "Anti"-Bayesian methods have also been used to design novel Prototype Reduction Schemes (PRS) [21] and new novel Border Identification (BI) algorithms [20]. The use of such "Anti"-Bayesian PRS and BI techniques in TC are extremely promising and are still unreported.

[5] As mentioned earlier, the authors of [17], [9] and [18] (cited in their chronological order) had initially proposed their results as being based on the *Order*-Statistics of the distributions. This was later corrected in [19], where they showed that their results were, rather, based on their *Quantile* Statistics.

[6] All of the theoretical results of [17], [9] and [18] were confirmed with rigorous experimental testing. The results of [18] were also proven on real-life data sets.

[7] In all the cases, they worked with the assumption that the a priori distributions were identical.

3.2   TC: A Multi-dimensional "Anti"-Bayesian Problem

Any problem that deals with TC must operate in a space that is very high dimensional primarily the because cardinality of the BOW can be very large. This, in and of itself, complicates the QS-based paradigm. Indeed, since we are speaking about the quantile statistics of a distribution, it implicitly and explicitly assumes that the points can be *ordered*. Consequently, the multi-dimensional generalization of CMQS, theoretically and with regard to implementation, is particularly non-trivial because there is no well-established method for achieving the ordering of multi-dimensional data specified in terms of its uni-dimensional components.

To clarify this, consider two patterns, $\mathbf{x}_1 = [x_{11}, x_{12}]^T = [2, 3]^T$ and $\mathbf{x}_2 = [x_{21}, x_{22}]^T = [1, 4]^T$. If we only considered the first dimension, x_{21} would be the first QS since $x_{11} > x_{21}$. However, if we observe the second component of the patterns, we can see that x_{12} would be the first QS. It is thus, clearly, not possible to obtain the ordering of the *vectorial* representation of the patterns based on their individual components, which is the fundamental issue to be resolved before the problem can be tackled in any satisfactory manner for multi-dimensional features. One can only imagine how much more complex this issue is in the TC domain – when the number of elements in the BOW is of the order of hundreds or even thousands.

To resolve this, multi-dimensional CQMS operates with a paradigm that is analogous to a Naïve-Bayes' approach, although it, really, is of an *Anti*-Naïve-Bayes' paradigm. Using such a *Anti*-Naïve-Bayes' approach, one can design and implement a CMQS-based classifier. The details of this design and implementation for two and multi-dimensions (and the associated conclusive experimental results) have been given in [18]. Indeed, on a deeper examination of these results, one will appreciate the fact that the higher-dimensional results for the various distributions do not necessarily follow as a consequence of the lower uni-dimensional results. They hold by virtue of the factorizability of the multi-dimensional density functions that follow the *Anti*-Naïve-Bayes' paradigm, and the fact that the d-dimensional QS-based statistics are concurrently used for the classification in every dimension.

3.3   Design and Implementation: "Anti"-Bayesian TC Solution

We shall now describe the design and implementation of the "Anti"-Bayesian TC solution.

"Anti"-Bayesian TC Solution: The Features. Each class is represented by two BOW vectors, one for each CMQS point used. For each class, we compute the frequency distribution of each word in each document in that class, and generate a frequency histogram for that word. While the traditional BOW approach would then pick the average value of this histogram, our method computes the area of the histogram and determines the two symmetric QS points. Thus, for example, if we are considering the $\frac{2}{7}$ and $\frac{5}{7}$ QS points of the two distributions, we would

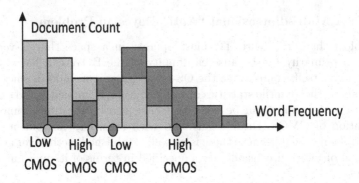

Fig. 1. Example of the QS-based features extracted from the histogram of a lower class (light grey) and of a higher class (dark grey), and the corresponding lower and higher CMQS points of each class.

pick the word frequencies that encompass the $\frac{2}{7}$ and $\frac{5}{7}$ of the histogram area respectively. The reader must observe the salient characteristic of this strategy: By working with such a methodology, for each word in the BOW, we represent the class by two of its non-central cases, rather than its average/median sample. This renders the strategy to be "Anti"-Bayesian!

For further clarity, we refer the reader to Fig. 1. For any word, the histograms of the two classes are depicted in light grey for the lower class, and in dark grey for the higher class. The QS-based features for the classes are then extracted from the histograms as clarified in the figure.

"Anti"-Bayesian TC Solution: The Multi-Class TC Classifier. Let us assume that the PR problem involves C classes. Since the "Anti"-Bayesian technique has been extensively studied for two-class problems, our newly-proposed multi-class TC classifier operates by invoking a sequence of $C-1$ pairwise classifiers. More explicitly, whenever a document for testing is presented, the system invokes a classifier that involves a pair of classes from which it determines a winning class. This winning class is then compared to another class until all the classes have been considered. The final winning class is the overall best and is the one to which the testing document is assigned.

"Anti"-Bayesian TC Solution: Testing. To classify an unknown document, we compute the cosine similarity between it and the features representing pairs of classes. This is done as follows: For each word, we mark one of the two groups as the high-group and the other as the low-group based on the word's frequency in the documents of each class, and we take the high CMQS point of the low-group and the low CMQS point of the high-group, as illustrated in Fig. 1. We build the two class vectors from these CMQS points, and we compute the cosine similarity between the document to classify each class vector using Eq. (1).

$$sim(c,d) = \frac{\sum_{i=0}^{W-1} w_{ic}w_{id}}{\sqrt{\sum_{i=0}^{W-1} w_{ic}^2}\sqrt{\sum_{i=0}^{W-1} w_{id}^2}}. \tag{1}$$

The most similar class is retained and the least similar one is discarded and replaced by one of the other classes to be considered, and the test is run again, until all the classes have been exhausted. The final class will be the most similar one, and the one that the document is classified into.

4   Experimental Set-Up

4.1   The Data Sets

For our experiments, we used the 20-Newsgroups corpus, a standard corpus in the literature pertaining to Natural Language Processing. This corpus contains 1,000 postings collected from the 20 different Usenet groups, each associated with a distinct topic, as listed in Table 1. We preprocessed each posting by removing header data (for example, "from", "subject", "date", etc.) and lines quoted from previous messages being responded to (which start with a '>' character), performing stop-word removal and word stemming, and deleting the postings that became empty of text after these preprocessing phases.

Table 1. The topics from the "20-Newsgroups" used in the experiments.

comp.graphics	alt.atheism	sci.crypt	misc.forsale
comp.sys.mac.hardware	talk.religion.misc	sci.electronics	rec.autos
comp.windows.x	talk.politics.guns	sci.med	rec.motorcycles
comp.os.ms-windows.misc	talk.politics.mideast	sci.space	rec.sport.hockey
comp.sys.ibm.pc.hardware	talk.politics.misc	soc.religion.christian	rec.sport.baseball

In every independent run, we randomly selected 70 % of the postings of each newsgroup to be used as training data, and retained the remaining 30 % as testing data.

4.2   The Histograms/Features Used

We first describe the process involved in the construction of the histograms and the extraction of the Quantile-based features.

Each document in the 20-Newsgroups dataset was preprocessed by word stemming using the Porter Stemmer algorithm and by a stopword removal phase. It was then converted to a BOW representation. The documents were then randomly assigned into training or testing sets.

The word-based histograms (please see Fig. 2) were then computed for each word in each category by tallying the observed frequencies for that word in each training document in that category, where the area of each histogram was the total sum of all the columns. The CMQS points were determined as those points where the cumulative sum of each column was equal to the CMQS moments when normalized with the total area. For further clarification, we present an example of two histograms[8] in Fig. 2 below. The $\frac{1}{3}$ and $\frac{2}{3}$ QS points of each histogram are marked along their horizontal axes. In this case, the markings represent the word frequencies that encompass the $\frac{1}{3}$ and $\frac{2}{3}$ areas of the histograms respectively. The histogram on the left depicts a less significant word for its category while the histogram on the right depicts a more significant word for its category. Note that in both histograms the first CMQS point is located at unity. To help clarify the figure, we mention that for the word "internet" in "rec.sport.baseball", both the CMQS points lie at unity - i.e., they are on top of each other.

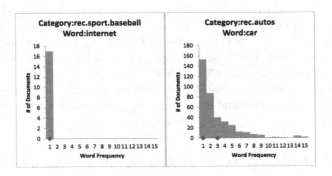

Fig. 2. The histograms and the $\frac{1}{3}$ and $\frac{2}{3}$ QS points for the two words "internet" and "car" from the categories "rec.sport.baseball" and "rec.autos".

4.3   The Benchmarks Used

We have developed three benchmarks for our system: A BOW classifier which involved the TFs and invoked the cosine similarity measure given by Eq. (1), a BOW classifier with the TFIDF features, and a Naïve-Bayes' classifier.

To understand how they all fit together, we define the Term Frequency (TF) of a word (synonymous with "term") t in a document d as Freq(t, d), and for each document this is calculated as the frequency count of the term in the document. This is, quite simply, given by Eq. (2):

$$\mathrm{TF}(t, d) = \mathrm{Freq}(t, d), \tag{2}$$

where Freq(t, d) is the number of times that the term t occurs in the document d.

---

[8] The documents used in this test were very short, which explains why the histograms are heavily skewed in favour of lower word frequencies.

The BOW classifier computes an average word/term vector $\mathbf{w_c}$ for each class c, which contains the average occurrence frequency of each of the W terms in that class (i.e., w_{tc}). It computes this by adding together the frequency count of each term as it occurs in each document of a class, and by then dividing the total by the number of documents in the class (N_c), as per Eq. (3).

$$w_{tc} = \frac{1}{N_c} \sum_{d=1}^{N_c} \text{TF}(t,d). \tag{3}$$

The quantity w_{tc} defined in Eq. (3) can also be seen to be the TF value as calculated per *class* instead of per document. Thus, to be explicit:

$$\text{TF}(t,c) = w_{tc}, \tag{4}$$

where w_{tc} is specified in Eq. (3).

Classifying a test document, d', is done by computing the cosine similarity of that test document's TF vector (which will likewise contain the occurrence frequency of each word in that document, $\text{TF}(t,d')$) with the TF for each each class, $\text{TF}(t,c)$, as per Eq. (1), and assigning the document to the most similar class.

The IDF, or Inverse Document Frequency, is the inverse ratio of the number of term vectors in the training corpus containing a given word. Specifically, if N_t is the number of classes in the training corpus containing a given term t, and C is the total number of classes in the corpus, the $\text{IDF}(t)$ is given as in Eq. (5):

$$\text{IDF}(t) = \log_{10} \frac{C}{N_t}. \tag{5}$$

Combining the above, we get the TFIDF value *per document* as the quantity calculated by:

$$\text{TFIDF}(t,d) = \text{TF}(t,d) \times \text{IDF}(t), \tag{6}$$

where $\text{TF}(t,d)$ is given by Eq. (2).

Analogously, the TFIDF value *per class* is the quantity calculated as:

$$\text{TFIDF}(t,c) = \text{TF}(t,c) \times \text{IDF}(t), \tag{7}$$

where $\text{TF}(t,c)$ is specified in Eq. (4).

The Naïve-Bayes' classifier selects the class c^* which is most probable one given the observed document, following Eq. (8). This is based on the prior probability of the class being independent of any other information, $P(c)$, multiplied by the probability of observing each individual word of the document t in the class, $P(t|c)$. This probability is computed as the frequency count of each word in the class divided by its frequency count in the entire corpus of N documents, as in Eq. (9). Finally, in order to avoid multiplications by zero in the case of a term that was never before seen in a class, we set the minimal value for $P(t|c)$ to be one thousandth of the minimum probability that was actually observed.

$$c^* = \arg\max_c \left[P(c) \prod_{t \in c} P(t|c) \right].$$ (8)

Also, since every class in the corpus had an equal number of documents and equal likelihood, the term for the *a priori* probability $P(c)$ in Eq. (8) was set to be always equal to 1/20, and was thus ignored.

$$P(w_i|c) = \frac{\sum_{d=1}^{N_c} w_{id}}{\sum_{d=1}^{N} w_{id}}.$$ (9)

4.4   The Testing and Accuracy Metrics Used

The Metrics Used. In every testing case, we used the respective data to train and test our classifier and each of the three benchmark schemes. For each newsgroup i, we counted the number TP_i of postings correctly identified by a classifier as belonging to that group, the number FN_i of postings that should have belonged in that group but were misidentified as belonging to another group, and the number FP_i of postings that belonged to other groups but were misidentified as belonging to this one. The Precision P_i is the proportion of postings assigned in group i that are correctly identified, and the Recall R_i is the proportion of postings belonging in the group that were correctly recognized, and are given by Eqs. (10) and (11) respectively. The F score is an average of these two metrics for each group, and the *macro-F1* is the average of the F scores over the all groups, and these are given in Eqs. (12) and (13) respectively.

$$P_i = \frac{TP_i}{TP_i + FP_i}$$ (10)

$$R_i = \frac{TP_i}{TP_i + FN_i}$$ (11)

$$F_i = \frac{2P_i R_i}{P_i + R_i}$$ (12)

$$macro\text{-}F1 = \frac{1}{20} \sum_{i=1}^{20} F_i$$ (13)

Correlation Between the Classifiers. Since the features and methods used in the classification are rather distinct, it would be a remarkable discovery if we could confirm that the results between the various classifiers are not correlated. In this regard, it is crucial to understand what the term "correlation" actually means. Formalized rigorously, the statistical correlation between two classifiers, X and Y would be defined as in Eq. (14) below:

$$\text{ClassifierCorr}_{X,Y} = \frac{\sum\limits_{i=1}^{N-1} \left(x_i - \bar{x}\right)\left(y_i - \bar{y}\right)}{N\sigma_X \sigma_Y}, \qquad (14)$$

where X and Y are the classifiers being compared, x_i and y_i are '0' or '1', and are the assigned values for incorrect and correct classifications of document i by X and Y respectively, $\bar{x}$ and $\bar{y}$ are the average performances of X and Y over all the documents, N is the number of documents, and σ_X and σ_Y are the standard deviations of the performances of X and Y respectively.

However, on a deeper examination, one would observe that while Eq. (14) yields the *statistical* correlation, it is only suited to classifiers that yield accuracies within the interval $[0, 1]$. It is, thus, not the best equation to compare the classifiers that we are dealing with. Rather, since the classifiers themselves yield binary results ('0' or '1' for incorrect or correct classifications), it is more appropriate to compare classifiers X and Y by the "number" of times they yield *identical* decisions. In other words, a more suitable metric for evaluating how any two classifiers X and Y yield identical results is given by Eq. (15) below:

$$\text{ClassifierSim}_{X,Y} = \frac{Pos_X Pos_Y + Neg_X Neg_Y}{Pos_X Pos_Y + Pos_X Neg_Y + Neg_X Pos_Y + Neg_X Neg_Y}, \qquad (15)$$

where $Pos_X Pos_Y$ and $Neg_X Neg_Y$ are the count of cases where the classifiers X and Y both return identical decisions '1' or '0' respectively, and where '0' and '1' represent the events of a classifier classifying a document incorrectly or correctly respectively. Analogously, $Pos_X Neg_Y$ and $Neg_X Pos_Y$ are the counts of cases where X returns '1' and Y returns '0' and vice-versa respectively. The reader should observe that strictly speaking, this metric would not yield a statistical correlation between the classifiers X and Y, but rather a statistical measure of their relative similarities. However, in the interest of maintaining a relatively acceptable terminology (and since we have previously used the term "similarity" to imply the similarity between *documents and classes* as opposed to the similarity between the *classifiers*), we shall informally refer to this classifier similarity as their mutual correlation, because, it does, in one sense, inform us about how correlated the decision made by classifier X is to the decision made by classifier Y.

5   Experimental Results

In this section, we shall present the results that we have obtained by testing our "Anti"-Bayesian (indicated, in the interest of brevity, by AB in the tables and figures) methodology against the benchmark classifiers described above. There are, indeed, two sets of results that are available: The first involves the case when the "Anti"-Bayesian scheme uses only the TF criteria, and this is done in Sect. 5.1. This is followed by the results when the "Anti"-Bayesian paradigm invokes the TFIDF criteria, i.e., when the lengths of the documents are also involved in characterizing the features. These results are presented in Sect. 5.2. A comparison and the correlation between these two sets of "Anti"-Bayesian schemes themselves is finally given in Sect. 5.3.

5.1 The Results Obtained: "Anti"-Bayesian TF Scheme

The experimental results that we have obtained for the "Anti"-Bayesian scheme that used only the TF criteria are briefly described below. We performed 100 tests, each one using a different random 70%/30% split of training and testing documents. We then evaluated the results of each classifier by computing the Precision, Recall, and F-score of each newsgroup, whence we computed the *macro-F1* value for each classifier over the 20-Newsgroups. The average results we obtained, over all 100 tests, are summarized in Table 2.

Table 2. The *macro-F1* score results for the 100 classifications attempted and for the different methods. In the case of the "Anti"-Bayesian scheme, the method used the TF features.

Classifier	CMQS points	*macro-F1* score
"Anti"-Bayesian	1/2, 1/2	0.709
	1/3, 2/3	0.662
	1/4, 3/4	0.561
	1/5, 4/5	0.465
	2/5, 3/5	0.700
	1/6, 5/6	0.389
	1/7, 6/7	0.339
	2/7, 5/7	0.611
	3/7, 4/7	0.710
	1/8, 7/8	0.288
	3/8, 5/8	0.686
	1/9, 8/9	0.264
	2/9, 7/9	0.515
	4/9, 5/9	0.713
	1/10, 9/10	0.243
	3/10, 7/10	0.631
BOW		0.604
BOW-TFIDF		0.769
Naïve-Bayes		0.780

We summarize the results that we have obtained:

1. The results show that for *half* of the CMQS pairs, the "Anti"-Bayesian classifier performed as well as and sometimes even better than the traditional BOW classifier. For example, while the BOW had a Macro-*F1* score of 0.604, the corresponding index for the CQMS pairs $\langle \frac{1}{3}, \frac{2}{3} \rangle$, was remarkably higher, i.e., 0.662. Further, the *macro-F1* score indices for $\langle \frac{2}{5}, \frac{3}{5} \rangle$, $\langle \frac{3}{7}, \frac{4}{7} \rangle$ and $\langle \frac{4}{9}, \frac{5}{9} \rangle$ were consistently higher – 0.700, 0.710 and 0.713 respectively. This, in itself, is quite remarkable, since our methodology is reversed to the traditional ones. This is also quite fascinating, given that it uses points distant from the mean (i.e., moving towards the extremities of the distributions) rather than the averages that are traditionally considered.

2. While the results obtained for extreme CMQS points very distant from the mean were not so impressive[9], the corresponding results for other non-central QS pairs were very encouraging. For example, the corresponding index for the CQMS pairs $\langle \frac{2}{7}, \frac{5}{7} \rangle$ was much higher than the BOW index, i.e., 0.611.

3. The results of the BOW and the "Anti"-Bayesian classifier were always less than what was obtained by the BOW-TFIDF and the Naïve-Bayes' classifier. This result is actually easily explained, because while all the classifiers compare vectors using cosine similarities, the BOW-TFIDF uses the more-informed document-weighted features. We shall presently show that if we use corresponding TFIDF-based features (that are more suitable for such text-based classifiers) with an "Anti"-Bayesian paradigm, we can obtain a comparable accuracy. That being said, the question of determining the best metric to be used for an "Anti"-Bayesian classifier in this syntactic space is currently unresolved.

Since the features/methodology used by the "Anti"-Bayesian classifier are different than those used by the traditional classifiers, it follows that they would perform differently, and either correctly or incorrectly classify different documents, as seen from a correlation-based analysis below. To verify this, we computed the correlation, as defined by Eq. (15), between the results of the "Anti"-Bayesian classifier in each of our 100 tests and the three benchmarks classifiers. Observe that a correlation near to unity would indicate that the corresponding two classifiers make identical decisions on the same documents – either correctly and incorrectly, while a correlation around '0' would indicate that their classification results are unrelated. The average correlation scores for the classifiers over all 100 tests are given in Table 3. The following points are noteworthy:

1. "Anti"-Bayesian classifiers that use CMQS points that are farther from the mean or median of the distributions show a lower correlation with the $\langle \frac{1}{2}, \frac{1}{2} \rangle$ "Anti"-Bayesian classifier. This is, actually, quite remarkable, considering that they sometimes give comparable accuracies even though they use *completely different features*. It also implies that two classifiers built from the same

---

[9] Given that these extreme points give better results in the next experiment when we classify using the TFIDF criteria (instead of merely the TF criteria), we hypothesize that this poor behavior is probably due to noise from non-significant words that is somehow amplified in the extreme CMQS points. But this issue is still unresolved.

Table 3. The correlation between the different classifiers for the 100 classifications achieved. In the case of the "Anti"-Bayesian scheme, the method used the TF features.

Classifier	CMQS points	AB at (1/2, 1/2)	BOW	BOW with TFIDF	Naïve-Bayes
"Anti"-Bayesian	1/2, 1/2	1.000	0.648	0.759	0.810
	1/3, 2/3	0.845	0.642	0.722	0.772
	1/4, 3/4	0.738	0.625	0.646	0.676
	1/5, 4/5	0.646	0.595	0.570	0.589
	2/5, 3/5	0.902	0.643	0.747	0.806
	1/6, 5/6	0.579	0.568	0.514	0.526
	1/7, 6/7	0.537	0.549	0.478	0.487
	2/7, 5/7	0.790	0.635	0.684	0.723
	3/7, 4/7	0.925	0.643	0.755	0.816
	1/8, 7/8	0.496	0.527	0.439	0.445
	3/8, 5/8	0.882	0.642	0.738	0.794
	1/9, 8/9	0.478	0.517	0.423	0.429
	2/9, 7/9	0.695	0.613	0.612	0.637
	4/9, 5/9	0.938	0.643	0.757	0.818
	1/10, 9/10	0.462	0.509	0.408	0.414
	3/10, 7/10	0.811	0.638	0.699	0.743
BOW		0.648	1.000	0.714	0.654
BOW-TFIDF		0.759	0.714	1.000	0.800
Naïve-Bayes		0.810	0.654	0.800	1.000

data and statistics but that utilize different CMQS points will have different behaviours and also yield different results. This is all the more interesting since, from Table 2, we can see that these classifiers will, in many cases, have similar *macro-F1* scores. This indicates that a fusion classifier that combines the information from multiple CMQS points could outperform any single classifier, and be built without requiring any additional data or tools from that classifier.

2. It is surprising to see that the "Anti"-Bayesian classifiers, almost consistently, have higher correlations with the two benchmarks that performed better than it. Indeed, the BOW-TFIDF classifier and the Naïve-Bayes' classifier show much larger correlations than the BOW classifier. In fact, the correlation between our "Anti"-Bayesian classifier and the BOW classifier is, almost always, the lowest of all the pairs, indicating that they generate the most different classification results!

3. Figure 3 displays the plots of the correlation between the different classifiers for the 100 classifications achieved, where in the case of the "Anti"-Bayesian scheme, the method used the TF features. The reader should observe the uncorrelated nature of the classifiers when the CMQS points are non-central, and the fact that this correlation increases as the feature points become closer to the mean or median.

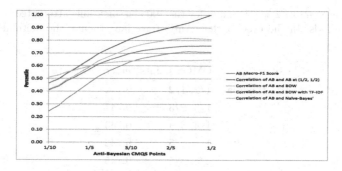

Fig. 3. Plots of the correlation between the different classifiers for the 100 classifications achieved. In the case of the "Anti"-Bayesian scheme, the method used the TF features.

5.2   The Results Obtained: "Anti"-Bayesian TFIDF Scheme

The results of the "Anti"-Bayesian scheme when it involves TFIDF features are shown in Table 4. In this case, the TF is calculated per document as per Eq. (6) for the test document, and as per Eq. (7) for each of the classes it is tested against. From this table we can glean the following results:

1. The results show that for *all* CMQS pairs, the "Anti"-Bayesian classifier performed much better than the traditional BOW classifier. For example, while the BOW had a *macro-F1* score of 0.604, the corresponding index for the CQMS pairs $\langle \frac{1}{3}, \frac{2}{3} \rangle$, was significantly higher, i.e., 0.747. Further, the *macro-F1* score indices for $\langle \frac{1}{4}, \frac{3}{4} \rangle$, $\langle \frac{3}{7}, \frac{4}{7} \rangle$ and $\langle \frac{4}{9}, \frac{5}{9} \rangle$ were consistently higher – 0.746, 0.744 and 0.744 respectively. This demonstrates the validity of our counter-intuitive paradigm – that we can truly get a remarkable accuracy even though we are characterizing the documents by the syntactic features of the points quite distant from the mean and more towards the extremities of the distributions.
2. In all the cases, the values of the Macro-*F1* index was only slightly less than the indices obtained using the BOW-TFIDF and the Naïve-Bayes approaches.

Since the features/methodology used by the "Anti"-Bayesian classifier are different than those used by the traditional classifiers, it is again advantageous to embark on a correlation-based analysis. To achieve this, we have again computed the correlation, as defined by Eq. (15) between the results of the "Anti"-Bayesian classifier (using the TFIDF criteria) in each of our 100 tests, and the three benchmarks classifiers. As before, a correlation near to unity would indicate that the corresponding two classifiers make identical decisions on the same documents – either correctly and incorrectly, while a correlation around '0' would indicate that their classification results are unrelated. The average correlation scores for the classifiers over all 100 tests are given in Table 5.

Table 4. The *macro-F1* score results for the 100 classifications attempted and for the different methods. In the case of the "Anti"-Bayesian scheme, the method used the TFIDF features.

Classifier	CMQS points	*macro-F1* score
"Anti"-Bayesian	1/2, 1/2	0.742
	1/3, 2/3	0.747
	1/4, 3/4	0.746
	1/5, 4/5	0.742
	2/5, 3/5	0.745
	1/6, 5/6	0.736
	1/7, 6/7	0.729
	2/7, 5/7	0.747
	3/7, 4/7	0.744
	1/8, 7/8	0.720
	3/8, 5/8	0.746
	1/9, 8/9	0.712
	2/9, 7/9	0.745
	4/9, 5/9	0.744
	1/10, 9/10	0.705
	3/10, 7/10	0.748
BOW		0.604
BOW-TFIDF		0.769
Naïve-Bayes		0.780

From the table, we observe the following rather remarkable points:

1. The first result that we can infer is that just as in the case when we used the TF features, the "Anti"-Bayesian classifier using the TFIDF criteria, when it works with CMQS points that are not near the mean or the median, has lower correlation than the benchmark classifiers that works with CMQS points that are near the mean or median, Indeed, they sometimes give comparable accuracies even though they use *completely different features*.

2. Again, the "Anti"-Bayesian classifier actually has the highest correlation in its results with the two benchmarks that performed better than it. This means that although the classification algorithm is similar to a BOW classifier, its results are more closely aligned to those of the more-informed TFIDF and NB classifiers.

3. Even when the "Anti"-Bayesian classifier used points very distant from the mean (for example, $\langle \frac{1}{10}, \frac{9}{10} \rangle$), the correlation was as high as 0.764. This means that there were more than 76 % of the cases when they both used completely different classifying criteria and yet produced similar results.

Table 5. The correlation between the different classifiers for the 100 classifications achieved. In the case of the "Anti"-Bayesian scheme, the method used the TFIDF features.

Classifier	CMQS points	AB at (1/2, 1/2)	BOW	BOW with TFIDF	Naïve-Bayes
"Anti"-Bayesian	1/2, 1/2	1.000	0.636	0.780	0.832
	1/3, 2/3	0.946	0.635	0.784	0.836
	1/4, 3/4	0.928	0.635	0.786	0.831
	1/5, 4/5	0.913	0.634	0.785	0.824
	2/5, 3/5	0.960	0.635	0.780	0.835
	1/6, 5/6	0.898	0.632	0.781	0.817
	1/7, 6/7	0.887	0.631	0.779	0.811
	2/7, 5/7	0.936	0.635	0.785	0.833
	3/7, 4/7	0.968	0.635	0.779	0.834
	1/8, 7/8	0.873	0.626	0.771	0.800
	3/8, 5/8	0.954	0.635	0.781	0.835
	1/9, 8/9	0.862	0.625	0.768	0.794
	2/9, 7/9	0.920	0.635	0.786	0.829
	4/9, 5/9	0.974	0.636	0.779	0.834
	1/10, 9/10	0.853	0.624	0.764	0.788
	3/10, 7/10	0.939	0.636	0.785	0.834
BOW		0.636	1.000	0.714	0.654
BOW-TFIDF		0.780	0.714	1.000	0.800
Naïve-Bayes		0.832	0.654	0.800	1.000

4. Figure 4 displays the plots of the correlation between the different classifiers for the 100 classifications achieved, where in the case of the "Anti"-Bayesian scheme, the method used the TFIDF features. The reader should observe the uncorrelated nature of the classifiers when the CMQS points are non-central. This correlation increases as the feature points become closer to the mean or median.

5.3 Correlation Between "Anti"-Bayesian TF *Versus* TFIDF Schemes

The correlated/uncorrelated nature of the "Anti"-Bayesian TF and TFIDF schemes with the other methods was explained in the earlier sections. It would be educative to examine how uncorrelated the "Anti"-Bayesian TF and the "Anti"-Bayesian TFIDF schemes are between themselves. In other words, even though their accuracies may be comparable, it would be good to examine if the two "Anti"-Bayesian classifiers are relatively uncorrelated in and of themselves. Thus, if a particular pair of CMQS points yielded distinct classification decisions using the two schemes, and if they, all the same, yielded comparable accuracies, the potential of the paradigm is shown to be significantly more. This is precisely

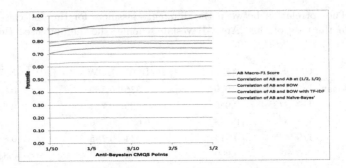

Fig. 4. Plots of the correlation between the different classifiers for the 100 classifications achieved. In the case of the "Anti"-Bayesian scheme, the method used the TFIDF features.

what we embark on achieving now – i.e., examining the correlation (or lack thereof) of the "Anti"-Bayesian TF and TFIDF schemes.

Table 6 reports the correlation, as defined by Eq. (15) between the results of the "Anti"-Bayesian classifier TF and TFIDF criteria in each of our 100 tests. The table also include the corresponding Macro-F1 scores. Again, a correlation near to unity would indicate that the two classifiers make identical decisions on the same documents – either correctly and incorrectly, while a correlation around '0' would indicate that their classification results are unrelated. The results tabulated in Table 6 are also depicted graphically in Fig. 5 whence the trends in the correlation with the increasing values of the CMQS points is clear.

From Table 6, we observe that:

1. When the CMQS points are close to the mean or median, the correlation is quite high (for example, 0.842). This is not surprising at all, since in such cases, the "Anti"-Bayesian classifier reduces to become a Bayesian classifier.
2. When the CMQS points are far from the mean or median, the correlation is quite high (for example, 0.659 for the CMQS points $\langle \frac{2}{9}, \frac{7}{9} \rangle$). This is quite surprising because although both schemes are "Anti"-Bayesian in their philosophy, the lengths of the documents play a part in determining the decisions that they individually make because the IDF values account for document lengths.
3. From the values of the associated Macro-F1 scores, we see that a lower correlation between these two classifiers is directly related to their difference in accuracy. This means that when the accuracies of the two classifiers are lower, each of them is classifying the documents on distinct criteria – which is far from being obvious.

This naturally leads us to our final section which deals with how we can fuse the results of the various classifiers.

On Utilizing Classifier Fusion. This section briefly touches on possible exploratory work, where we consider how the various classifiers can be "fused".

Table 6. The correlation between the two "Anti"-Bayesian classifiers for the 100 classifications when they utilized the TF and the TFIDF features respectively.

Classifier	CMQS points	AB Macro-F1	AB TFIDF Macro-F1	Correlation of AB and AB TFIDF
"Anti"-Bayesian	1/2, 1/2	0.709	0.742	0.842
	1/3, 2/3	0.662	0.747	0.792
	1/4, 3/4	0.561	0.746	0.699
	1/5, 4/5	0.465	0.742	0.616
	2/5, 3/5	0.700	0.745	0.833
	1/6, 5/6	0.389	0.736	0.557
	1/7, 6/7	0.339	0.729	0.523
	2/7, 5/7	0.611	0.747	0.745
	3/7, 4/7	0.710	0.744	0.845
	1/8, 7/8	0.288	0.720	0.493
	3/8, 5/8	0.686	0.746	0.819
	1/9, 8/9	0.264	0.712	0.481
	2/9, 7/9	0.515	0.745	0.659
	4/9, 5/9	0.713	0.744	0.848
	1/10, 9/10	0.243	0.705	0.472
	3/10, 7/10	0.631	0.748	0.762

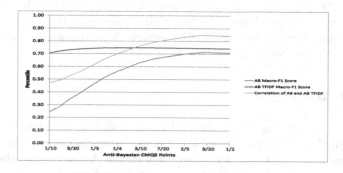

Fig. 5. The correlation between the two "Anti"-Bayesian classifiers for the 100 classifications when they utilized the TF and the TFIDF features respectively.

Combined with the aforementioned fact that they use a completely different set of features for classification, and that they are the two simplest of the five classifiers we considered, let us consider how the BOW and "Anti"-Bayesian scheme using the TF features can be fused. Indeed, it would be interesting to see how they could be combined by incorporating a relatively simple data fusion technique. As a preliminary *prima facie* experiment in that direction, we combined the classification of the BOW classifier and our "Anti"-Bayes classifier (using the TF criteria) in each of our 100 experiments. Since each classifiers measures the similarity between a document and the classes' feature vectors and then picks the maximum, we performed this combination simply by comparing the winning (for example, the highest) class similarity value returned by each of the two classifiers and picking the maximum one. We found that this classifier obtains an average *macro-F1* score of 0.674, only marginally better than the 0.671 *macro-F1* score of the best "Anti"-Bayes classifier in our tests. Upon further examination, we find that this is due to the fact that the similarity values generated by the "Anti"-Bayes classifier are on average three times higher than those generated by the BOW classifier. Consequently, the "Anti"-Bayes classification is the one picked in almost all cases! However, the few cases where the BOW classifier's similarity score beats that of the "Anti"-Bayes classifier are also cases where the BOW correctly classified documents that the "Anti"-Bayes classifier missed, leading to the small improvement observed in the results. Moreover, our data shows that there are more than 1,000 documents (over 12 % of the test corpus) that the BOW classifier correctly classifies with a similarity that is less than that of the "Anti"-Bayesian's erroneous classification.

There is thus clear room for improvements in the final classification, and the main challenge for future research will involve developing a fair weighting scheme between the two classifiers in order to compensate for the lower similarity scores of the BOW classifier, without misclassifying the over 1,500 test documents that the "Anti"-Bayesian classifier recognizes correctly but that the BOW misclassifies.

Indeed, the potential of designing fused classifiers involving the BOW, the BOW-TFIDF, the Naïve Bayes, the "Anti"-Bayesian using the TF criteria, and the "Anti"-Bayesian that uses the TDIDF criteria, is extremely great considering their relative accuracies and correlations.

6   Conclusions

In this paper we have considered the problem of Text Classification (TC), which is a problem that has been studied for decades. From the perspective of classification, problems in TC are particularly fascinating because while the feature extraction process involves *syntactic or semantic* indicators, the classification uses the principles of *statistical* Pattern Recognition (PR). The state-of-the-art in TC uses these statistical features in conjunction with the well-established methods such as the Bayesian, the Naïve Bayesian, the SVM etc. Recent research has advanced the field of PR by working with the Quantile Statistics (QS) of the features. The resultant scheme called Classification by Moments of Quantile Statistics (CMQS) is essentially "Anti"-Bayesian in its *modus operandus*, and advantageously works with information latent in "outliers" (i.e., those distant from the mean) of the distributions. Our goal in this paper was to demonstrate the power and potential of CMQS to work within the *very* high-dimensional TC-related vector spaces and their "non-central" quantiles. To investigate this, we considered the cases when the "Anti"-Bayesian methodology used both the TD and the TFIDF criteria.

Our PR solution for C categories involved $C-1$ pairwise CMQS classifiers. By a rigorous testing on the well-acclaimed data set involving the 20-Newsgroups corpus, we demonstrated that the CMQS-based TC attains accuracy that is comparable to and sometimes even better than the BOW-based classifier, even though it essentially uses the information found only in the "non-central" quantiles. The accuracies obtained are comparable to those provided by the BOW-TFIDF and the Naïve Bayes classifier too!

Our results also show that the results we have obtained are often uncorrelated with the established ones, thus yielding the potential of fusing the results of a CMQS-based methodology with those obtained from a more traditional scheme.

References

1. Alahmadi, A., Joorabchi, A., Mahdi, A.E.: A new text representation scheme combining bag-of-words and bag-of-concepts approaches for automatic text classification. In: Proceedings of the 7th IEEE GCC Conference and Exhibition, Doha, Qatar, pp. 108–113, November 2014
2. Debole, F., Sebastiani, F.: Supervised term weighting for automated text categorization. In: Proceedings of the 18th ACM Symposium on Applied Computing, Melbourne USA, pp. 784–788, March 2003
3. Duda, R.O., Hart, P.E., Stork, D.G.: Pattern Classification. A Wiley Interscience Publication, New York (2006)
4. Dumoulin, J.: Smoothing of n-gram language models of human chats. In: Proceedings of the Joint 6th International Conference on Soft Computing and Intelligent Systems (SCIS) and 13th International Symposium on Advanced Intelligent Systems (ISIS), Kobe, Japan, pp. 1–4, November 2012
5. Lu, L., Liu, Y.-S.: Research of english text classification methods based on semantic meaning. In: Proceedings of the ITI 3rd International Conference on Information and Communications Technology, Cairo, Egypt, pp. 689–700, December 2005
6. Madsen, R.E., Sigurdsson, S., Hansen, L.K., Larsen, J.: Pruning the vocabulary for better context recognition. In: Proceedings of the 17th International Conference on Pattern Recognition, Cambridge, UK, vol. 2, pp. 483–488, August 2004
7. Menon, R., Keerthi, S.S., Loh, H.T., Brombacher, A.C.: On the effectiveness of latent semantic analysis for the categorization of call centre records. In: Proceedings of the IEEE International Engineering Management Conference, Singapore, vol. 2, pp. 545–550 (2004)
8. Ning, Y., Zhu, T., Wang, Y.: Affective-word based chinese text sentiment classification. In: Proceedings of the 5th International Conference on Pervasive Computing and Applications (ICPCA), Maribor, Slovenia, pp. 111–115, December 2010
9. Oommen, B.J., Thomas, A.: Optimal order statistics-based "Anti-Bayesian" parametric pattern classification for the exponential family. Pattern Recogn. **47**, 40–55 (2014)
10. Ouamour, S., Sayoud, H.: Authorship attribution of ancient texts written by ten arabic travelers using character N-Grams. In: Proceedings of the 2013 International Conference on Computer, Information and Telecommunication Systems (CITS), Piraeus-Athens, Greece, pp. 1–5, May 2013

11. Qiang, G.: An effective algorithm for improving the performance of Naïve Bayes for text classification. In: Proceedings of the Second International Conference on Computer Research and Development, Kuala Lumpur, Malaysia, pp. 699–701, May 2010

12. Salton, G., McGill, M.: Introduction to Modern Information Retrieval. Mc-Graw Hill Book Company, New York (1983)

13. Salton, G., Wong, A., Yang, C.S.: A vector space model for automatic indexing. Comm. ACM **18**, 613–620 (1975)

14. Salton, G., Yang, C.S., Yu, C.: A theory of term importance in automatic text analysis. Technical report, Ithaca, NY, USA (1974)

15. Salton, G., Yang, C.S., Yu, C.: Term weighting approaches in automatic text retrieval. Technical report, Ithaca, NY, USA (1987)

16. Sebastiani, F.: Machine learning in automated text categorization. ACM Comput. Surv. **34**, 1–47 (2002)

17. Thomas, A., Oommen, B.J.: The fundamental theory of optimal "Anti-Bayesian" parametric pattern classification using order statistics criteria. Pattern Recogn. **46**, 376–388 (2013)

18. Thomas, A., Oommen, B.J.: Order statistics-based parametric classification for multi-dimensional distributions. Pattern Recogn. **46**, 3472–3482 (2013)

19. Thomas, A., Oommen, B.J.: Corrigendum to three papers that deal with "Anti"-Bayesian pattern recognition. Pattern Recogn. **47**, 2301–2302 (2014)

20. Thomas, A., Oommen, B.J.: A novel border identification algorithm based on an "Anti-Bayesian" paradigm. In: Proceedings of CAIP'13, the 2013 International Conference on Computer Analysis of Images and Patterns, York, UK, pp. 196–203, August 2013

21. Thomas, A., Oommen, B.J.: Ultimate order statistics-based prototype reduction schemes. In: Proceedings of AI 2013, The 2013 Australasian Joint Conference on Artificial Intelligence, Dunedin, New Zealand, pp. 421–433, December 2013

22. Wu, G., Liu, K.: Research on text classification algorithm by combining statistical and ontology methods. In: Proceedings of the International Conference on Computational Intelligence and Software Engineering, Wuhan, China, pp. 1–4, December 2009

Two Novel Techniques to Improve MDL-Based Semi-Supervised Classification of Time Series

Vo Thanh Vinh[1]([✉]) and Duong Tuan Anh[2]

[1] Faculty of Information Technology, Ton Duc Thang University,
Ho Chi Minh City, Vietnam
vtvinh@it.tdt.edu.vn
[2] Faculty of Computer Science and Engineering, Ho Chi Minh City University
of Technology, Ho Chi Minh City, Vietnam
dtanh@cse.hcmut.edu.vn

Abstract. Semi-supervised classification problem arises in the situation that we just have a small amount of labeled instances in the training set. One method to classify the new time series in such situation is that; firstly we need to use self-training to classify the unlabeled instances in the training set. Then, we use the output training set to classify the new time series. In this paper, we propose two novel improvements for Minimum Description Length-based semi-supervised classification of time series: an improvement technique for Minimum Description Length-based stopping criterion and a refinement step to make the classifier more accurate. Our first improvement applies the non-linear alignment between two time series when we compute Reduced Description Length of one time series exploiting the information from the other. The second improvement is a post-processing step that aims to identify the class boundary between positive and negative instances accurately. For the second improvement, we propose an algorithm called Refinement that attempts to identify the wrongly classified instances in the self-training step; then it reclassifies these instances. We compare our method with some previous methods. Experimental results show that our two improvements can construct more accurate semi-supervised time series classifiers.

Keywords: Time series · Semi-supervised classification · Stopping criterion · MDL principle · X-Means

1 Introduction

In time series data mining, classification is a crucial problem which has attracted lots of research works in the last decade. However, most of the current methods assume that the training set contains a great number of labeled data. Such an assumption is unrealistic in the real world where we have a small set of labeled data, in addition to abundant unlabeled data. In such circumstances, semi-supervised classification is a suitable paradigm.

To the best of our knowledge, most of the studies about semi-supervised classification of time series follow two directions: the first approach bases on Wei and Keogh

© Springer-Verlag GmbH Germany 2016
N.T. Nguyen et al. (Eds.): TCCI XXV, LNCS 9990, pp. 127–147, 2016.
DOI: 10.1007/978-3-662-53580-6_8

framework [8] as in [1, 6, 8], and the second approach bases on a clustering algorithm such as in [10–12].

For the former approach, Semi-supervised classification (SSC) method will train itself by trying to expand the set of labeled data with the most similar unlabeled data until reaching a stopping criterion. Though several semi-supervised approaches have been proposed, only a few could be used for time series data, due to its special characteristic within. Most of the time series SSC methods have to suggest a good stopping criterion. The SSC approach for time series proposed by Wei et al. in 2006 [8] uses a stopping criterion which is based on the minimal nearest neighbor distance, but this criterion can not work correctly in some situations. Ratanamahatana and Wanichsan, in 2008 [6], proposed a stopping criterion for SSC of time series which is based on the historical distances between candidate instances from the set of unlabeled instances to the initial positive instances. The most well-known stopping criterion so far is the one using Minimum Description Length (MDL) proposed by Begum et al., 2013 [1]. Even though this newest state-of-the-art stopping criterion gives a breakthrough for SSC of time series, it is still not effective to be used in some situations where time series may have some distortion along the time axis and the computation of Reduced Description Length for them becomes so rigid that the stopping point for the classifier can not be found precisely.

For the latter approach, Nhut et al. in 2011 proposed a method called LCLC (Learning from Common Local Cluster) [11]. This method firstly apply K-means clustering algorithm to obtain the clusters. Then, it considers all the instances in a cluster belong to a class. According to Begum et al. [1], this method depends too much on the clustering algorithm and it wrongly classifies many instances. In order to improve LCLC, Nhut et al. in 2012 [12] proposed an extended version of LCLC called En-LCLC (Ensemble based Learning from Common Local Clusters). This method attempts to identify probability that a time series belong to a class. Since, the authors proposed a fuzzy classification algorithm called AFNN (Adaptive Fuzzy Nearest Neighbor) based on these probabilities. According to Begum et al. [1], this method needs to be set up many initial constants. Marussy and Buza in 2013 [10] proposed a semi-supervised classification method based on single-link hierarchical clustering accompanying with must-link constraint and cannot-link constraint. Different from the other methods, Marussy and Buza applied graph theory to tackle the semi-supervised classification problem. In this method, the authors showed that semi-supervised classification problem is equivalent to finding the minimal spanning tree problem in a graph. However, this method required to know all the classes before hand. For example, in binary classification, we need to classify into two classes. Marussy and Buza's method requires that there must be two types of instances labeled positive and negative as seeds at the beginning whereas the other methods only require one type of instances (positive instances only).

In this work, we propose two novel improvements for binary SSC of time series in the spirit of the first approach direction: an improvement technique for MDL-based stopping criterion and a refinement step to make the classifier more accurate. Our first improvement applies the non-linear alignment between two time series when we compute Reduced Description Length of one time series exploiting the information from the other. In order to obtain the non-linear alignment between two time series, we apply the Dynamic Time Warping distance. For the second improvement, we propose a post-processing step that aims to identify the class boundary between positive and negative instances accurately. Experimental results reveal that our two improvements can construct more accurate semi-supervised time series classifiers.

The rest of this paper is organized as follows. Section 2 reviews some background. Section 3 gives details of the two proposed improvements, followed by a set of experiments in Sect. 4. Section 5 concludes the work and gives suggestions for future work. Section Appendix shows some more experimental results.

2 Background

In this section, we review briefly Time Series and 1-Nearest Neighbor Classifier, Euclidean Distance, Dynamic Time Warping, and the framework of semi-supervised time series classification as well as some stopping criterion such as MDL-based stopping criterion, Ratanamahatana and Wanichsan's Stopping Criterion, and lastly we introduce X-means clustering algorithm.

2.1 Time Series and 1-Nearest Neighbor Classifier

A time series T is a sequence of real numbers collected at regular intervals over a period of time: $T = t_1, t_2, ..., t_n$. Furthermore, a time series can be seen as an n-dimensional object in metric space. In 1-Nearest Neighbor Classifier (1-NN), the data object is classified the same class as its nearest object in the training set. The 1-NN has been considered hard to beat in classification of time series data among many other methods such as Artificial Neural Network, Bayesian Network [16].

2.2 Euclidean Distance

The Euclidean Distance (ED) between two time series $Q = q_1, q_2, ..., q_n$ and $C = c_1, c_2, ..., c_n$ is a similarity measure defined as follows:

$$ED(Q, C) = \sqrt{\sum_{i=1}^{n} (q_i - c_i)^2}$$

Euclidean distance is one of the most widely used distance measure in time series, its computational complexity is $O(n)$. In this work, Euclidean Distance is applied only in the X-means clustering algorithm which is used to support the Refinement process described in Subsect. 3.2.

2.3 Dynamic Time Warping Distance

One problem with time series data is the distortion in the time axis, making Euclidean distance unsuitable. However, this problem can be effectively addressed by Dynamic

Time Warping (DTW), a distance measure that allows non-linear alignment between the two time series to accommodate sequences that are similar in shape but out of phase [2].

Now we would like to show how to calculate DTW. Given two time series Q and C which have length n and m respectively: $Q = q_1, q_2 \ldots, q_n$ and $C = c_1, c_2 \ldots, c_m$. DTW is a dynamic programming technique which calculates all possible warping paths between two time series for finding minimum distance. To calculate DTW between the two above time series, firstly we construct a matrix D with size $m \times n$. Every element in matrix D is cumulative distance defined as:

$$\gamma(i,j) = d(i,j) + \min \begin{cases} \gamma(i-1,j) \\ \gamma(i,j-1) \\ \gamma(i-1,j-1) \end{cases}$$

where $\gamma(i, j)$ is (i, j) element of matrix that is a summation between $d(i, j) = (q_i - c_j)^2$, a square distance of q_i and c_j, and the minimum cumulative distance of three adjacent elements to (i, j).

Next, we choose the optimal warping path which has minimum cumulative distance defined as:

$$DTW(Q, C) = \min \sum_{k=1}^{K} w_k$$

where w_k is (i, j) at k^{th} element of the warping path, and K is the length of the warping path.

In addition, for a more accurate distance measure, some global constraints were suggested to DTW. A well-known constraint is Sakoe-Chiba band [7], shown in Fig. 1. The Sakoe-Chiba band constrains the indices of the warping path $w_k = (i, j)_k$ such that $j - r \leq i \leq j + r$, where r is a term defining the allowed range of warping, for a given point in a sequence. Much more detail about DTW is beyond the scope of this paper, interested readers may refer to [3, 7].

Due to evident advantages of DTW for time series data, we incorporate DTW distance measure into our proposed algorithm.

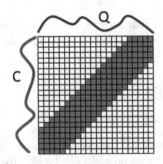

Fig. 1. DTW with Sakoe-Chiba band

2.4 Semi-Supervised Classification of Time Series

SSC technique can help build better classifiers in situations where we have a small set of labeled data, in addition to abundant unlabeled data. The main ideas of SSC of time series are summarized as follows. Given a set P of positive instances and a set N of unlabeled instances, the algorithm iterates the following two steps:

- *Step* 1: We find the nearest neighbor of any instance of our training set from the unlabeled instances.
- *Step* 2: This nearest neighbor instance, along with its newly acquired positive label, will be added into the training set.

Note that the above algorithm has to be coupled with the ability to stop adding instances at the correct time. This important issue will be addressed later. The algorithm for SSC of time series [1, 8] is given as follows:

Self_Training_Classifier (P, N)
    // P: Positive/Labeled set and N: Negative/Unlabeled set
while (the stopping criterion)
        nearest_obj = One_Nearest_Neighbor (P, N)
        $P = P \cup$ {nearest_obj}
        $N = N \setminus$ {nearest_obj}
End

Figure 2 illustrates the Semi-Supervised Learning process. The circled instances are the initial positive/labeled instances. The triangle instances are the positive/unlabeled instances, and the rectangle instances are the negative/unlabeled instances. Initially, there are three positive labeled instances (circled instances); the process will assign all the other unlabeled instances as well as their newly acquired labels into the positive set. As we can see, the positive/unlabeled will be added into the training set in a chain which is called the chain effect of this algorithm.

In this semi-supervised classification framework, to identify the point where negative instances are taken into the positive set is an important task as it affects the quality of the final training set. There are some stopping criterions were proposed such as

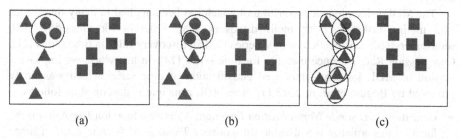

(a)   (b)   (c)

Fig. 2. Semi-Supervised Learning on time series data, (a) Initial positive/labeled instances (circled instances), (b) Select one nearest neighbor from unlabeled data (triangle instance) to added in to positive/labeled set, (c) Continue taking more unlabeled instances into positive/labeled set

Ratanamahatana and Wanichsan's Stopping Criterion [6] and Stopping Criterion based on MDL Principle [1], which are depicted in the next two subsections.

2.5   Ratanamahatana and Wanichsan's Stopping Criterion

In 2008, Ratanamahatana and Wanichsan [6] proposed a stopping criterion called SCC (Stopping Criterion Confidence) for semi-supervised classification of time series data which is based on the following formula:

$$SCC(i) = \frac{|Mindist(i) - Mindist(i - 1)|}{Std\{Mindist(1), Mindist(2), \ldots, Mindist(i)\}} \\ \times \frac{NumInitialUnlabeled - (i - 1)}{NumInitialUnlabeled}$$

- *Mindist*: minimum distance in the positive/labeled set after each step of adding one more instance into positive/labeled set.
- *Std*: standard deviation.
- *NumInitialUnlabeled*: the number of unlabeled data at the beginning of the learning phase.

At the point, the value of *SCC* is maximal, i.e. at iteration *i*, the stopping criterion is chose at $i - 2$.

In this work, we use this stopping criterion in order to test the effect of our Refinement process (described later in Subsect. 3.2) for Semi-Supervised Learning.

2.6   Stopping Criterion Based on MDL Principle

The Minimum Description Length (MDL) principle is a formalization of Occam's razor in which the best hypothesis for a given set of data is the one that leads to the best compression of the data. The MDL principle was introduced by Rissanen in 1978 [17]. This principle is a crucial concept in information theory and computational learning theory.

The MDL principle is a powerful tool which has been applied in many time series data mining tasks, such as motif discovery [18], criterion for clustering [19], semi-supervised classification of time series [1, 15], discovery rules in time series [21], Compression Rate Distance measure for time series [14]. In this work, we improve a version of MDL for semi-supervised classification of time series which was firstly proposed by Begum et al, in 2013 [1]. The MDL principle is described as follows:

- **Definition 1.** *Discrete Normalization Function*: A discrete function *Dis_Norm* is the function to normalize a real-value subsequence T into b-bit discrete value of range $[1, 2^b]$. The maximum of the discrete range value 2^b is also called the *cardinality*. The *Dis_Norm* function is described as follows:

$$Dis\_Norm(T) = round\left(\frac{T - \min}{\max - \min} \times (2^b - 1)\right) + 1$$

where *min* and *max* are the minimum and maximum value in T respectively.
After casting the original real-valued data to discrete values, we are interested in determining how many bits are needed to store a particular time series T. It is called the *Description Length* of T.

- **Definition 2.** *Description Length*: A description length DL of a time series T is the total number of bits required to represent it.

$$DL(T) = w \times \log_2 c$$

where w is the length of T and c is the cardinality (the number of values we discretize the time series).

- **Definition 3.** *Hypothesis*: A hypothesis H is a subsequence used to encode one or more subsequences of the same length.

We are interested in how many bits are required to encode T given H. It is called the *Reduced Description Length* of T.

- **Definition 4.** *Reduced Description Length*: A reduced description length of a time series T given hypothesis H is the sum of the number of bits required in order to encode T exploiting the information in H. i.e. $DL(T \mid H)$, and the number of bits required for H itself, i.e. $DL(H)$. Thus, the reduced description length is defined as:

$$DL(T, H) = DL(H) + DL(T|H)$$

One simple approach of encoding T using H is to store a *difference vector* between T and H. Therefore: $DL(T \mid H) = DL(T - H)$.

Example: Given A and H, two time series of length 20 as follows:

$$A = [6\,7\,9\,9\,10\ 11\,13\,13\,14\,15\ 16\,18\,18\,19\,22\,21\,22\,23\,24\,24]$$
$$H = [6\,7\,8\,9\,10\ 11\,12\,13\,14\,15\ 16\,17\,18\,19\,20\ 21\,22\,23\,24\,25]$$

Without encoding, the bit requirement to store A and H is $2 \times 20 \times \log_2 20 = 173$ bits. The difference vector $A' = |A - H| = [0\,0\,1\,0\,0\,0\,1\,0\,0\,0\,0\,1\,0\,0\,2\,0\,0\,0\,0\,1]$. And in the difference vector, there are 5 mismatches. The bit requirement is now just $20 \times \log_2 20 + 5 \times (\log_2 20 + \lceil \log_2 20 \rceil) = 134$ bits, which brings out a good data compression.

Assume that there exists only a single positive instance as the initial training set [1]. The SSC procedure using MDL-based stopping criterion can be outlined as follows.

First, it selects the seed positive instance as hypothesis. It selects the nearest neighbor of any of the instance(s) in the labeled training set from the unlabeled dataset. It encodes this instance in terms of the hypothesis and keeps the rest of the dataset uncompressed. Then it computes the reduced description length of the whole dataset. If it can achieve a data compression then the instance in question is a true positive. It continues to test to see if unlabeled instances can be added to the positive pool by this data compression criterion. Once the SSC module starts including instances dissimilar to the hypothesis, it no longer achieves data compression and the first occurrence of such an instance is the point where the SSC module should *stop*.

Even though this stopping criterion is the best one for SSC of time series so far, it is still not effective to be used in some situations where time series may have some distortion along the time axis and the way of computing Difference Vector for them becomes so rigid that the stopping point for the classifier can not be found precisely.

In this work, we improve this stopping criterion by applying a non-linear alignment between two time series when calculating their Reduce Description Length (described in Subsect. 3.1).

2.7   X-Means Clustering Algorithm

X-means was proposed by Pelleg and Moore in 2000 [5], which is an extended clustering algorithm of K-means. X-means can identify the best number of clusters k by itself based on the Bayesian Information Criterion (BIC) [20]. This clustering algorithm requires setting up a more flexible k cluster than in K-means. At the beginning, we need to specify a maximal value max_k and minimal value min_k of k clusters. X-means will identify which value of k in the range [min_k, max_k] should be selected. In Fig. 3, we show the outline of X-means which includes two steps. Step 1, called *Improve-Params*, runs K-means until converging. Step 2, called *Improve-Structure*, decides whether a cluster should be split into two sub-clusters or not basing on BIC. The algorithm stops when the number of clusters reaches the maximum number of cluster max_k which was set at the beginning.

In this work, we use X-means as a semi-supervised classification method, called *X-means-classifier*. We apply X-means-classifier to support our refinement step to identify the ambiguous instances which will be depicted later in Subsect. 3.2. For more information about X-means algorithm, interested reader can refer to [5].

	X-means
1	*Improve-Params*
2	*Improve-Structure*
3	If $K > K_{max}$, return the best-scoring model. Otherwise, go to step 1.

Fig. 3. Outline of X-means clustering algorithm [5]

3   The Proposed Method

This work aims to improve the MDL-based stopping criterion and at the same time improve the accuracy of the classifier. We devise an improvement technique for the MDL-based stopping criterion and propose a Refinement step to make the classifier more accurate.

3.1 New Stopping Criterion Based on MDL Principle

The original MDL-based stopping criterion is really simple, which finds mismatch points by one-to-one alignment between two time series and then calculates Reduced Description Length using the number of mismatch points. In fact, it is hard to find bit saves in this method because the time series may have some distortion in the time axis and a lot of mismatches will be found and there are not many bit saves.

We propose a more flexible technique for finding mismatch points. Instead of linear alignment, we use a non-linear alignment when finding mismatch points. This method attempts to find an optimal matching between two time series for determining as fewer mismatch points as possible.

The principle of our proposed method is in the same spirit of the main characteristic of Dynamic Time Warping (DTW). Therefore, we can modify the algorithm of computing DTW distance between two time series in order to include the finding of mismatch points between them.

Given an example, suppose we have two discrete time series H and A as follows:

$$H = [2\,6\,6\,8\,5]$$
$$A = [1\,6\,8\,5\,4]$$

By original method, the number of mismatch points is 4 because they have different values at 4 positions (2 vs. 1, 6 vs. 8, 8 vs. 5, and 5 vs. 4). On the other hand, by using our *Count_Mismatch* algorithm, the number of mismatch points is 2, less than in the original method. This result can be easily seen in Fig. 4. The alignment between A and H is shown in Fig. 4(a) through the warping path and the number of mismatch points between them is shown in Fig. 4(b).

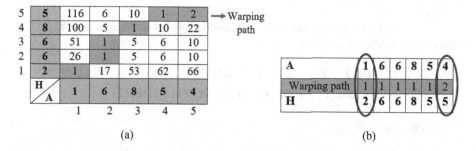

(a)   (b)

Fig. 4. Example of counting mismatch points in our proposed method

Figure 6 shows our proposed mismatch count algorithm based on the calculation of DTW distance. There are two phases in this algorithm. At first phase, we calculate the DTW distance. The second phase goes backward along the found warping path and finds the number of mismatch points. In addition, at first phase, we use Sakoe-Chiba band constraint (through the user-specified parameter r) for limiting the meaningless warping paths between the two time series.

In addition, for finding an efficient warping path, we also propose a method to calculate the suitable value of Sakoe-Chiba band r with the algorithm given in Fig. 5. At the beginning, the positive/labeled set must have at least *two* time series. We will calculate the value of r by finding the lowest value of r that satisfies the condition whereby one time series (seed) will accept the other as a positive instance. This condition results in the following inequality that must be satisfied:

$$mismatch\_count \leq TS\_length \times \frac{\log_2 card}{\log_2 card + \lceil \log_2 TS\_length \rceil}$$

where *mismatch_count* is the number of mismatch points between two positive/labeled time series, *TS_length* is the length of two time series and *card* is the cardinality.

r = Find_Match_Range ($T1$, $T2$, $card$)	
// $T1$, $T2$: positive/labeled sample time series,	
// $card$: the cardinality	
// TS_length: the length of two time series	
1	value = $TS\_length \times \log_2(card)/(\log_2(card) + \text{ceil}(\log_2(TS\_length)))$
2	**for** $i = 0$ **to** TS_length
3	mismatch_count = Count_Mismatch ($T1$, $T2$, i)
4	**if** mismatch_count <= value
5	**break**
6	**end**
7	**end**
8	$r = i$

Fig. 5. The outline of Refinement process in SSC

Now we will prove the above-mentioned condition.

Proof:

Time series $T1$ accept time series $T2$ as a positive/labeled instance, if and only if the following inequality is satisfied:

$$DL(T1, T2) \leq DL(T1) + DL(T2)$$
$$\Leftrightarrow DL(T1) + DL(T2 | T1) \leq DL(T1) + DL(T2)$$

We can derive:

$$DL(T2 | T1) \leq DL(T2)$$
$$\Leftrightarrow mismatch\_count \times (\log_2 card + \lceil \log_2 TS\_length \rceil) \leq TS\_length \times \log_2 card$$

So we can rewrite:

$$mismatch\_count \leq TS\_length \times \log_2 card/(\log_2 card + \lceil \log_2 TS\_length \rceil) \qquad \square$$

```
mismatch_count = Count_Mismatch (x, y, r)
   // x: Time series, y: Time series, r: Sakoe-Chiba band constraint
          // Phase 1: Calculate DTW with Sakoe-Chiba band constraint
  1       matrix[1,1] = square(x[1] – y[1])
  2       for i = 2 to length(y) do
  3          matrix[1, i] = matrix[1, i – 1] + square(x[1] – y[i])
  4       end
  5       for i = 2 to length(x) do
  6          matrix[i, 1] = matrix[i – 1, 1] + square (x[i] – y[1])
  7       end
  8       for i = 2 to length(x) do
  9          for j = 2 to length(y) do
 10            if |i – j| <= r then
 11               min_val = MIN(matrix[i – 1, j], matrix[i, j – 1], matrix[i – 1, j – 1])
 12               matrix[i, j] = min_val + square(x[i] – y[j])
 13            else
 14               matrix[i, j] = +INFINITY
 15            end
 16          end
 17       end
          // Phase 2: Finding minimum number of mismatch points
 18       i = length(x); j = length(y)
 19       mismatch_count = 0
 20       if x[i] != y[j] then
 21          mismatch_count = mismatch_count + 1
 22       end
 23       while i > 1 OR j > 1 do
 24          value = MIN(matrix[i – 1, j], matrix[i, j – 1], matrix[i – 1, j – 1])
 25          if i > 1 AND j > 1 AND value = matrix[i – 1, j – 1] then
 26             i = i – 1; j = j – 1
 27          else if j > 1 AND value = matrix[i, j – 1] then
 28             j = j – 1
 29          else if i > 1 AND value = matrix[i – 1, j] then
 30             i = i – 1
 31          end
 32          if x[i] != y[j] then
 33             mismatch_count = mismatch_count + 1
 34          end
 35       end
```

Fig. 6. Mismatch-count algorithm between two time series with Sakoe-Chiba band constraint

Based on the above inequality, we proposed the algorithm for finding the suitable value for the Sakoe-Chiba band r, which is given in Fig. 5. Line 3 of the algorithm in Fig. 5 invokes the procedure *Count_Mismatch* which is given in Fig. 6. This algorithm can be easily extended for finding r with more than two initial positive/labeled samples. One solution on this situation is to choose r as the average value of Match Range between any two pairs of positive/labeled time series.

3.2   Refinement Step

In this work, we include to the framework of semi-supervised time series classification algorithm given in Subsect. 3.2 a process called *Refinement*. The aim of this process is to check again the training set and modify it in order to obtain a more accuracy classifier. This process is based on the finding of *ambiguous labeled instances*, and these ambiguous instances will be classified again using the confident true labeled instances. The refinement process is iterated until the training set becomes stable, i.e. the training set before and after a refinement iteration are the same.

Figure 7 shows our proposed refinement algorithm. In this algorithm, *AMBI* is the set of ambiguous labeled instances, *P* is the positive set and *N* is the negative set. The set *AMBI* consists of the instances which are near the positive and negative boundary. This algorithm classifies the instances in *AMBI* basing on the current *P* and *N*. The process of detecting *AMBI* and classifying the instances in *P* is repeated until *P* and *N* are unchanged. Finally, the instances in *AMBI* that cannot be labeled will be classified the last time.

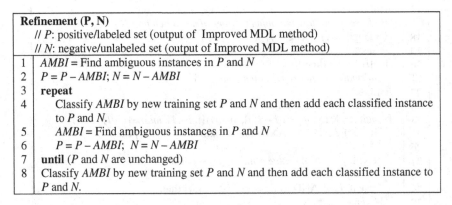

Refinement (P, N)
  // *P*: positive/labeled set (output of Improved MDL method)
  // *N*: negative/unlabeled set (output of Improved MDL method)

1	*AMBI* = Find ambiguous instances in *P* and *N*
2	*P* = *P* – *AMBI*; *N* = *N* – *AMBI*
3	**repeat**
4	Classify *AMBI* by new training set *P* and *N* and then add each classified instance to *P* and *N*.
5	*AMBI* = Find ambiguous instances in *P* and *N*
6	*P* = *P* – *AMBI*; *N* = *N* – *AMBI*
7	**until** (*P* and *N* are unchanged)
8	Classify *AMBI* by new training set *P* and *N* and then add each classified instance to *P* and *N*.

Fig. 7.   The outline of Refinement process in SSC

The ambiguous instance detection process is done under the following rules:

1. The instances in *P* which were classified as positive by SSC but their nearest neighbors are in the negative set *N*, they and their nearest neighbors are ambiguous.
2. The instances in *N* which were classified as negative by SSC but their nearest neighbors are in the positive set *P*, they and their nearest neighbors are ambiguous.
3. The instances which were classified as positive by *X-means-classifier* (explained later) but are classified as negative by SSC, these are considered ambiguous.

The process of classifying instances in *AMBI* is done using One-Nearest-Neighbor (1-NN) in which the instance in *AMBI* which is nearest to *P* or *N* will be labeled first.

In this work, we propose a method called *X-means-Classifier* that can be used as SSC method for time series. This is a clustering-based approach which applies *X-means* algorithm, an extended variant of *k-means* which was proposed by Pelleg and Moore in

2000 [5]. One outstanding feature of *X-means* is that it can automatically estimate the suitable number of clusters during the clustering process. The SSC method based on *X-means* consists of the following steps. First, we use *X-means* to cluster the training set (including positive and unlabeled instances). Then, if there exists one cluster which contains the positive instance, all the instances in it will be classified as positive instances, and all the rest are classified as negative. *X-means-Classifier* will be used to initialize the *AMBI* in the Refinement process (Line 1 in the algorithm in Fig. 7).

In Fig. 8, we show an example to illustrate how the Refinement process works. In Fig. 8(a), the circled/positive instances and squared/negative instances are obtained from the Self-Learning process. The separate line which split the space into two areas *P* and *N* indicates the true boundary between two classes *P* and *N*. As we can see from Fig. 8(a), there are three wrongly classified instances, two squared instances indicate that they belong to negative set but their true class is positive (they stand in area *P*), and one circled instance indicates that it belong to positive set but their true class is negative (because it locates in *N* area). When applying the Refinement process, some ambiguous instances are identified because their nearest neighbors belong to another class as shown in Fig. 8(b). Since, they are reclassified as shown in Fig. 8(c). In Fig. 8(d), the Refinement process is continued, two more instances are identified as ambiguous instances. They are finally reclassified as in Fig. 8(e). The Refinement step repeats until there is no change in the positive set and the negative set.

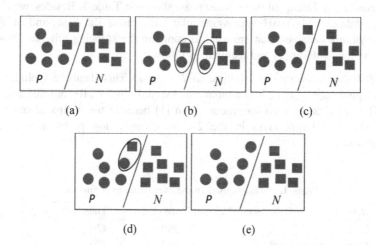

(a)   (b)   (c)

(d)   (e)

Fig. 8. An example of Refinement step, (a) positive set *P* and negative set *N* after applying Self-Learning with improved MDL-based stopping criterion, (b) ambiguous instances are identified (the two pair of instances marked), (c) the ambiguous instances are reclassified, (d) continuing to identify ambiguous instances, (e) the final training set after Refinement step

4 Experimental Evaluation

We implemented our proposed method and previous methods with Matlab 2012 and conducted the experiments on the Intel Core i7-740QM 1.73 GHz, 4 GB RAM PC. After the experiments, we evaluate the classifier by measuring the precision, recall and

F-measure of the retrieval. The precision is the ratio of the correctly classified positive test data to the total number of test instances classified as positive. The recall is the ratio of the correctly classified positive test data to the total number of all positive instances in the test dataset. An F-measure is the ratio defined by the formula:

$$F = \frac{2 \times p \times r}{p + r}$$

where p is precision and r is recall.

$$p = \frac{\# \ of \ correct \ positive \ predictions}{number \ of \ positive \ predictions}$$

$$r = \frac{\# \ of \ correct \ positive \ predictions}{number \ of \ positive \ examples}$$

In general, the higher the F-measure is, the better the classifier is.

4.1   Datasets

Our experiments were conducted over the datasets from UCR Time Series Classification Archive [4]. Details of these datasets are shown in Table 1. Besides, we also use two other datasets: MIT-BIH Supraventricular Arrhythmia Database, and St. Petersburg Arrhythmia Database that are used to compare the stopping criteria. These two datasets are available in [9] and featured as follows:

- *MIT-BIH Supraventricular Arrhythmia Database*: This database includes many ECG signals and a set of beat annotations by cardiologists. Record 801 and signal ECG1 were used in our experiments as in [1] because we compared our method with [1]. The target class in the 2-class classification problem is abnormal heartbeats.

Table 1. Datasets used in the evaluation experiments

Datasets	Number of classes	Size of dataset	Time series length
Yoga	2	300	426
Words synonyms	25	267	270
Two patterns	4	1000	128
MedicalImages	10	381	99
Synthetic control	6	300	60
TwoLeadECG	2	23	82
Gun-Point	7	50	150
Fish	7	175	463
Lightming-2	2	60	637
Symbols	6	25	398

- *St. Petersburg Arrhythmia Database*: This database contains 75 annotated readings extracted from 32 Holter records. Record I70 and signal II were used in our experiments as in [1] because we compared our method with [1]. The target class in the 2-class classification problem is R-on-T Premature Ventricular Contraction.

4.2 Parameters Setup

Cardinality for the MDL principle (described in Subsect. 2.6) is set to 8 (3-bit discrete values). For all the methods, we use DTW as distance measure. Euclidean Distance is applied only in X-means-classifier.

4.3 Comparing Two MDL-Based Stopping Criteria

We perform a comparison between our improvement technique and the previous MDL-based stopping criteria [1] on four datasets: MIT-BIH Supraventricular Arrhythmia Database, St. Petersburg Arrhythmia Database, Gun Point Training Set and Fish Training Set in Figs. 9, 10, 11 and 12 respectively. In order to compare the stopping criteria, we record the point when the truly negative instance is added into the positive set of Self-Learning process, this point is consider as expected stopping point. We compare the stopping criteria based on this expected stopping point as a baseline.

From Figs. 9, 10, 11 and 12, we can see that our improvement technique suggests a better stopping point in most of the datasets. Detecting a good stopping point is very crucial in SSC of time series. We attribute this desirable advantage of our improvement technique to the flexible way of determining mismatches between two time series when computing Reduced Description Length of one time series exploiting the information in the other.

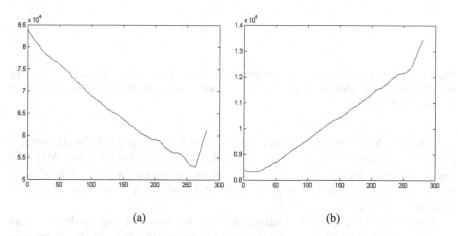

(a)   (b)

Fig. 9. In MIT-BIH Supraventricular Arrhythmia Database, the expected stopping point is 268. (a) Stopping point by our MDL (Proposed Method) at iteration 262 (Nearly perfect). (b) Stopping point by MDL (Previous Method) at iteration 10 (too early).

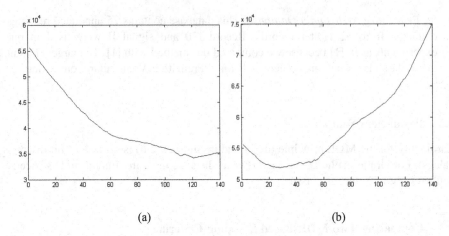

(a)   (b)

Fig. 10. In St. Petersburg Arrhythmia Database, the expected stopping point is 126. (a) Stopping point by our MDL (Proposed Method) at iteration 121 (Nearly perfect). (b) Stopping point by MDL (Previous Method) at iteration 28 (too early)

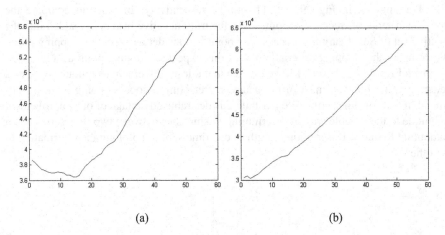

(a)   (b)

Fig. 11. In Gun Point Training Set, the expected stopping point is 14th. (a) Stopping point by our MDL (Proposed Method) at iteration 15th (Nearly perfect). (b) Stopping point by MDL at iteration 3rd (too early).

Figure 9 shows the experimental results of our proposed MDL based stopping criterion compared with the previous MDL based stopping criterion MIT-BIH Supraventricular Arrhythmia Database. Our proposed stopping point is 268 which is nearly the same as expected stopping point 262, and much better than that of the previous method at 10.

Figures 10, 11 and 12 also reveal that our improvement can produce a more accurate stopping point than the previous stopping criterion. In St. Petersburg Arrhythmia Database (Fig. 10), the expected stopping point is 126; our proposed method gives result 128, whereas the previous method gets 28 as stopping point.

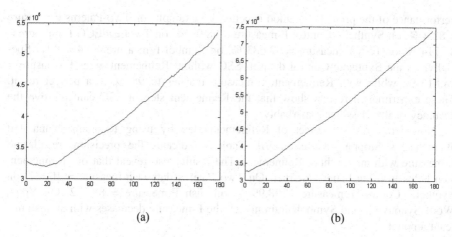

(a)   (b)

Fig. 12. In Fish Training Set, the expected stopping point is 18th. (a) Stopping point by our MDL (Proposed Method) at iteration 19th (Nearly perfect). (b) Stopping point by MDL at iteration 3rd (too early).

In Gun Point (Fig. 11), the expected stopping point is 14; our proposed method gives result 15, whereas the previous method gets 3 as stopping point. And in Fish dataset (Fig. 12), the expected stopping point is 18; our proposed method gives result 19, whereas the previous method gets 3 as stopping point.

4.4 Effects of Refinement Step

In this subsection, we compare SSC by our new MDL-based stopping criterion with and without Refinement step. Table 2 reports the experimental results (precision, recall and F-measure) of this comparison. The results show that our proposed Refinement step brings out better performance in all the datasets. In most of datasets, the

Table 2. Experiment results with and without Refinement (used proposed stopping criterion)

Datasets	Without Refinement			With Refinement		
	Precision	Recall	F-measure	Precision	Recall	F-measure
Yoga	0.64	0.35036	0.45283	0.57609	0.38686	**0.46288**
WordsSynonyms	0.94737	0.3	0.4557	0.625	0.41667	**0.5**
Two patterns	1.0	0.41328	0.58486	1.0	0.68635	**0.814**
MedicalImages	0.57276	0.91133	0.70342	0.56587	0.93103	**0.70391**
Synthetic control	1.0	0.08	0.14815	1.0	0.98	**0.9899**
TwoLeadECG	0.88889	0.66667	0.7619	0.75	1.0	**0.85714**
Gun-Point	0.93333	0.58333	0.71795	1.0	0.625	**0.76923**
Fish	0.94737	0.81818	0.87805	1.0	0.86364	**0.92683**
Lightning-2	0.7619	0.4	0.52459	0.6875	0.55	**0.61111**
Symbols	1.0	0.75	0.85714	0.88889	1.0	**0.94118**

performance of the proposed method is better, for example, on Two-Patterns F-measure = 81.4 %, on Synthetic-Control F-measure = 98.99 %, on TwoLeadECG F-measure = 85.714 %, on Fish F-measure = 92.683 %, on Symbol F-measure = 94.118 %. Specially, on the Synthetic-Control dataset, SSC without Refinement gives F-measure = 14.815 %, while with Refinement, F-measure reaches to 98.99 %, a perfect result. These experimental results show that the Refinement step in SSC can improve the accuracy of the classifier remarkably.

Now we show the effect of Refinement step by using Ratanamahatana and Wanichsan's Stopping Criterion [6]. Table 3 indicates the precision, recall and F-measure with and without Refinement. The results also reveal that our Refinement step helps to bring better classifier. On Two Paterns dataset F-measure = 100 %, on Synthetic Control F-measure = 98.99 %, on Fish F-measure = 88.372. On Yoga, WordsSynonyms, and Symbols training set, the F-measure decreases with an insignificant amount.

Table 3. Experiment results with and without Refinement (used Ratanamahatana and Wanichsan's Stopping Criterion [6])

Datasets	Without Refinement			With Refinement		
	Precision	Recall	F-measure	Precision	Recall	F-measure
Yoga	0.6383	0.43796	0.51948	0.58654	0.44526	0.50622
WordsSynonyms	0.58696	0.9	0.71053	0.45669	0.96667	0.62032
Two patterns	0.99267	1.0	0.99632	1.0	1.0	**1.0**
MedicalImages	0.96078	0.24138	0.38583	1.0	0.24138	**0.38889**
Synthetic control	0.95918	0.94	0.94949	1.0	0.98	**0.9899**
TwoLeadECG	1.0	0.41667	0.58824	0.71429	0.83333	**0.76923**
Gun-Point	0.63636	0.58333	0.6087	0.68182	0.625	**0.65217**
Fish	0.9	0.81818	0.85714	0.90476	0.86364	**0.88372**
Lightning-2	0.62162	0.575	0.5974	0.69388	0.85	**0.76404**
Symbols	0.53333	1.0	0.69565	0.5	1.0	0.66667

5   Conclusions

Existing semi-supervised learning algorithms for time series classification still have less than satisfactory performance. In this work, we have proposed two novel improvements for semi-supervised classification of time series: an improvement technique for MDL-based stopping criterion and a refinement step to make the classifier more accurate. Our former improvement applies the Dynamic Time Warping to find a non-linear alignment between two time series when computing their Reduced Description Length. The latter improvement attempts to identify wrongly classified instances by self-learning process and reclassify these instances. Experimental results reveal that our two improvements can construct more accurate semi-supervised time series classifiers.

As for future work, we plan to generalize our method to the case of multiple classes and adapt it to some other distance measures such as Complexity-Invariant Distance [13] or Compression Rate Distance [14]. Compression Rate Distance is a powerful distance measure for time series data which we recently proposed. We also plan to include some constraint in the Semi-Supervised Learning process as in [15] and extend our method to perform semi-supervised classification for streaming time series. Besides, we intend to apply another version of MDL such as in [14, 15, 19] which computes the Description Length of a time series by its entropy. Although our method helps to improve the F-measure of the output training set, there are still many instances which were wrongly classified in the training set. This weakness could be solved by removing the wrongly classified instances.

Acknowledgment. We would like to thank Prof. Eamonn Keogh and Nurjahan Begum for kindly sharing the datasets which help us in constructing the experiments in this work.

Appendix A: Some More Experimental Results

This section shows the experimental results of X-means-classifier which was used to support the Refinement step shown in Subsect. 4.4. Table 4 illustrates the precision, recall and F-measure of X-means classifier. The experiments reveal that X-means classifier gives good results in some datasets such as in Synthetic Control F-measure = 100 %, in Symbols F-measure = 94.118 %, in Gun Point F-measure = 71.795 %, in Fish F-measure = 71.795 %. Specially, in Synthetic Control, the result is perfect F-measure = 100 % (totally exact).

Table 4. Semi-supervised classification of time series by X-means-Classifier

Datasets	Precision	Recall	F-measure
Yoga	0.48421	0.33577	0.39655
WordsSynonyms	0.35632	0.51667	0.42177
Two patterns	0.28676	0.28782	0.28729
MedicalImages	0.71277	0.33005	0.45118
Synthetic control	1.0	1.0	1.0
TwoLeadECG	0.6	0.75	0.66667
Gun-Point	0.93333	0.58333	0.71795
Fish	0.82353	0.63636	0.71795
Lightning-2	0.75	0.525	0.61765
Symbols	0.88889	1.0	0.94118

In Table 5, we show the execution time (seconds) of some algorithms: Refinement with Improved MDL based stopping criterion, Refinement with Ratanamahatana and Wanichsan's stopping criterion, and X-means-classifier. Note that these figures do not include the execution time of Self-Learning process.

Table 5. The execution time (seconds) of each algorithm

Datasets	Improved MDL based stopping criterion with Refinement	Ratanamahatana and Wanichsan's stopping criterion with Refinement	X-means
Yoga	32.158997	5.398214	5.349535
WordsSynonyms	17.201075	2.3947555	2.313145
Two patterns	23.461983	10.6709155	7.588841
MedicalImages	5.4952995	1.2302655	1.124553
Synthetic control	1.8147455	0.623597	0.564281
TwoLeadECG	0.3780895	0.038229	0.030232
Gun-Point	0.4166765	0.1593195	0.147925
Fish	6.669053	2.237528	2.202647
Lightning-2	4.60413	0.777428	0.763867
Symbols	0.964603	0.2640985	0.258735

References

1. Begum, N., Hu, B., Rakthanmanon, T., Keogh, E.J.: Towards a minimum description length based stopping criterion for semi-supervised time series classification. In: IEEE 14th International Conference on Information Reuse and Integration, IRI 2013, San Francisco, CA, USA, August 14–16, 2013, pp. 333–340 (2013)
2. Berndt, D.J., Clifford, J.: Using dynamic time warping to find patterns in time series. In: Knowledge Discovery in Databases: Papers from the 1994 AAAI Workshop, Seattle, Washington, July 1994. Technical report WS-94-03, pp. 359–370 (1994)
3. Keogh, E.J., Ratanamahatana, C.A.: Exact indexing of dynamic time warping. Knowl. Inf. Syst. **7**(3), 358–386 (2005)
4. Chen, Y., Keogh, E., Hu, B., Begum, N., Bagnall, A., Mueen, A., Batista, G.: The UCR time series classification archive, July 2015. www.cs.ucr.edu/~eamonn/time_series_data/
5. Pelleg, D., Moore, A.W.: X-means: extending k-means with efficient estimation of the number of clusters. In: Proceedings of the Seventeenth International Conference on Machine Learning (ICML 2000), Stanford University, Stanford, CA, USA, June 29–July 2, 2000, pp. 727–734 (2000)
6. Ratanamahatana, C.A., Wanichsan, D.: Stopping criterion selection for efficient semi-supervised time series classification. In: Software Engineering, Artificial Intelligence, Networking and Parallel/Distributed Computing, pp. 1–14 (2008)
7. Sakoe, H., Chiba, S.: Dynamic programming algorithm optimization for spoken word recognition. IEEE Trans. Acoust. Speech Signal Process. **26**(1), 43–49 (1978)
8. Wei, L., Keogh, E.J.: Semi-supervised time series classification. In: Proceedings of the Twelfth ACM SIGKDD International Conference on Knowledge Discovery and Data Mining, Philadelphia, PA, USA, August 20–23, 2006, pp. 748–753 (2006)
9. Begum, N.: Minimum description length based stopping criterion for semi-supervised time series classification (2013). www.cs.ucr.edu/~nbegu001/SSL_myMDL.htm
10. Marussy, K., Buza, K.: SUCCESS: a new approach for semi-supervised classification of time-series. In: Rutkowski, L., Korytkowski, M., Scherer, R., Tadeusiewicz, R., Zadeh, L.A., Zurada, J.M. (eds.) ICAISC 2013, Part I. LNCS, vol. 7894, pp. 437–447. Springer, Heidelberg (2013). doi:10.1007/978-3-642-38658-9_39

11. Nguyen, M.N., Li, X.L., Ng, S.K.: Positive unlabeled learning for time series classification. In: Proceedings of the Twenty-Second International Joint Conference on Artificial Intelligence, IJCAI 2011, vol. 2, pp. 1421–1426. AAAI Press (2011)

12. Nguyen, M., Li, X.-L., Ng, S.-K.: Ensemble based positive unlabeled learning for time series classification. In: Lee, S., Peng, Z., Zhou, X., Moon, Y.-S., Unland, R., Yoo, J. (eds.) DASFAA 2012, Part I. LNCS, vol. 7238, pp. 243–257. Springer, Heidelberg (2012). doi:10. 1007/978-3-642-29038-1_19

13. Batista, G.E.A.P.A., Keogh, E.J., Tataw, O.M., de Souza, V.M.A.: CID: an efficient complexity-invariant distance for time series. Data Min. Knowl. Discov. 28(3), 634–669 (2014)

14. Vinh, V.T., Anh, D.T.: Compression rate distance measure for time series. In: 2015 IEEE International Conference on Data Science and Advanced Analytics, DSAA 2015, Campus des Cordeliers, Paris, France, October 19–21, 2015, pp. 1–10 (2015)

15. Vinh, V.T., Anh, D.T.: Constraint-based MDL principle for semi-supervised classification of time series. In: 2015 Seventh International Conference on Knowledge and Systems Engineering, KSE 2015, Ho Chi Minh City, Vietnam, October 8–10, 2015, pp. 43–48 (2015)

16. Ding, H., Trajcevski, G., Scheuermann, P., Wang, X., Keogh, E.J.: Querying and mining of time series data: experimental comparison of representations and distance measures. PVLDB 1(2), 1542–1552 (2008)

17. Rissanen, J.: Modeling by shortest data description. Automatica 14(5), 465–471 (1978)

18. Tanaka, Y., Iwamoto, K., Uehara, K.: Discovery of time-series motif from multidimensional data based on MDL principle. Mach. Learn. 58(2–3), 269–300 (2005)

19. Rakthanmanon, T., Keogh, E.J., Lonardi, S., Evans, S.: MDL-based time series clustering. Knowl. Inf. Syst. 33(2), 371–399 (2012)

20. Schwarz, G.E.: Estimating the dimension of a model. Ann. Stat. 6(2), 461–464 (1978)

21. Shokoohi-Yekta, M., Chen, Y., Campana, B.J.L., Hu, B., Zakaria, J., Keogh, E.J.: Discovery of meaningful rules in time series. In: Proceedings of the 21th ACM SIGKDD International Conference on Knowledge Discovery and Data Mining, Sydney, NSW, Australia, August 10–13, 2015, pp. 1085–1094 (2015)

Author Index

Printed in the United States
By Bookmasters